BIOCHEMICAL SOCIETY SYMPOSIA

No. 64

CELLULAR RESPONSES TO STRESS

BIOCHEMICAL SOCIETY SYMPOSIUM No. 64
held at The University of Liverpool, Spring 1997

Cellular Responses to Stress

ORGANIZED AND EDITED BY

C. P. DOWNES, C. R. WOLF AND D. P. LANE

PORTLAND PRESS

Published in the United Kingdom by Portland Press,
59 Portland Place, London W1N 3AJ, U.K.
Tel: (+44) 171 580 5530, e-mail: edit@portlandpress.co.uk

Published in North America by Princeton University Press,
41 William Street, Princeton, NJ 08540, U.S.A.

ISBN 1 85578 123 9 **ISSN 0067-8964**

British Library Cataloguing in Publication Data
A catalogue record for this book is available from the British Library

Typeset by Unicus Graphics Ltd, Horsham, U.K.
Printed in Great Britain by Information Press Ltd, Eynsham, U.K.

Contents

Preface vii

Abbreviations ix

1 Signal transduction by the c-Jun N-terminal kinase
By R.J. Davis 1

2 Roles of the AMP-activated/SNF1 protein kinase family in the response to cellular stress
By D.G. Hardie 13

3 Making the connection: coupling of stress-activated ERK/MAPK (extracellular-signal-regulated kinase/mitogen-activated protein kinase) core signalling modules to extracellular stimuli and biological responses
By J.M. Kyriakis 29

4 Stress-activated MAP kinase (mitogen-activated protein kinase) pathways of budding and fission yeasts
By J.B.A. Millar 49

5 Protein kinase cascades in intracellular signalling by interleukin-1 and tumour necrosis factor
By J. Saklatvala, J. Dean and A. Finch 63

6 Regulation of actin dynamics by stress-activated protein kinase 2 (SAPK2)-dependent phosphorylation of heat-shock protein of 27 kDa (Hsp27)
By J. Landry and J. Huot 79

7 DNA-dependent protein kinase and related proteins
By G.C.M. Smith, N. Divecha, N.D. Lakin and S.P. Jackson 91

8 Stress-induced activation of the heat-shock response: cell and molecular biology of heat-shock factors
By J.J. Cotto and R.I. Morimoto 105

9 Transcriptional regulation via redox-sensitive iron–sulphur centres in an oxidative stress response
By B. Demple, E. Hidalgo and H. Ding 119

10 Adaptive responses to environmental chemicals
By C.R. Wolf, G. Smith, A.G. Smith, K. Brown and C.J. Henderson 129

11 Cellular response to cancer chemopreventive agents: contribution of the antioxidant responsive element to the adaptive response to oxidative and chemical stress
By J.D. Hayes, E.M. Ellis, G.E. Neal, D.J. Harrison and M.M. Manson 141

Subject index 169

Preface

All cells have evolved complex signalling pathways which allow them to adapt to, and survive, toxic agents in their environment. These pathways are fundamental to the life and death of cells and determine their sensitivity to particular toxic agents. As such, cellular responses to stress play a central role in the sensitivity of micro-organisms to chemotherapeutic agents, and in the sensitivity of normal and tumour cells to environmental agents and anti-cancer drugs. Therefore an understanding of the pathways of cellular response to stress is fundamental to human health.

Environmental factors play a pivotal role in the pathogenesis of most human diseases, and when these adaptive responses fail to cope with the environment to which they are exposed, cell death and disease processes are instigated. On the other hand, the capacity of cells to adapt to chemical and DNA-damaging agents is critical to the success of chemotherapy, and the overexpression of many of these genes has been directly related to drug resistance.

This book summarizes the talks presented at a Biochemical Society Meeting held in July 1997. The aim of the meeting was to bring together a group of expert scientists who work on different but inter-related aspects of cellular stress responses. Although the meeting covered a wide range of different types of cellular stress in different organisms, it will be apparent that many of the adaptive responses involved are highly conserved through phylogeny and therefore studies on simple organisms are directly relevant to those observed in more complex multicellular eukaryotes.

It is the aim of this text to provide state-of-the-art information on the wide spectrum of ways in which cells can respond to different forms of stress induced by chemicals, oxidants and DNA-damaging agents. Mechanisms are described that involve altered uptake and efflux of chemical agents, intracellular detoxification and DNA damage responses. Many of these changes trigger a cascade of reactions mediated by stress-activated signalling pathways. These pathways have the capacity to determine whether a cell will survive or die. In this book, the spectrum of topics covered aims to provide a broad overview of our current knowledge of the different forms of adaptive response systems.

It is hoped that this text will stimulate further research to establish the relative cellular role of specific response pathways, and enable us to gain a deeper understanding of the pathways that allow cells to live or die.

C. P. Downes, C. R. Wolf and D. P. Lane

Abbreviations

AFAR	aflatoxin aldehyde reductase
AICAriboside	5-aminoimidazole-4-carboxamide riboside
AMPK	AMP-activated protein kinase
AMPKK	AMPK kinase
AP-1	activator protein-1
ARE	antioxidant responsive element
ASC	association with the SNF1 complex
ASK	apoptosis-stimulating kinase
A-T	ataxia–telangiectasia
ATF	activating transcription factor
ATM	A-T mutated
ATR	A-T-related
BHA	butylated hydroxyanisole
tBHQ	t-butylhydroquinone
bZIP	basic region leucine zipper
CHOP	CCAAT-enhancer-binding protein homologous protein
CREB	cAMP response element binding protein
CRIB	Cdc42/Rac interaction and binding
CSBP	cytokine-suppressive anti-inflammatory drug binding protein
CYP	cytochrome *P*-450
DMBA	dimethylbenzanthracene
DNA-PK	DNA-dependent protein kinase
DNA-PK_{cs}	catalytic subunit of DNA-PK
EGF	epidermal growth factor
eIF	eukaryotic initiation factor
EMSA	electrophoretic mobility shift assay
ERK	extracellular-signal-regulated kinase
FADD	Fas-associated death domain
FGF	fibroblast growth factor
FKBP12	FK506-binding protein
FRAP	FKBP12–rapamycin-associated protein
GCK	germinal centre kinase
GCS	γ-glutamylcysteine synthase
GCS_h	gene encoding the GCS heavy subunit
GEF	guanine nucleotide exchange factor
GST	glutathione S-transferase

HMG-CoA	3-hydroxy-3-methylglutaryl-CoA
HO-1	haem oxygenase-1
HPK	haematopoietic progenitor kinase
HR-C	heptad repeat C
HSF	heat-shock transcription factor
Hsp	heat-shock protein
HUVEC	human umbilical vein endothelial cells
IκB	inhibitor of NF-κB
IL-1	interleukin-1
JNK	c-Jun N-terminal kinase
Krs	kinase responsive to stress
Ku	DNA-binding component of DNA-PK
LPS	lipopolysaccharide
MAP	mitogen-activated protein
MAPK	MAP kinase
MAPKAPK	MAP-kinase-activated protein kinase
MAPKK	MAP kinase kinase
MAPKKK	MAP kinase kinase kinase
MEK	MAP-kinase/ERK kinase
MEKK	MKK kinase
MKK	MAP kinase kinase
MKKK	MKK kinase
MLK	mixed-lineage kinase
MnK	MAP kinase integrating kinase
MST	mammalian sterile twenty-like
NF-κ	nuclear factor κ
NIK	Nck-interacting kinase
NIK_{MAPKKK}	NF-κ-inducing kinase
NQO	NAD(P)H quinone oxidoreductase
PAK	p21-activated kinase
PI	phosphatidylinositol
PMA	phorbol 12-myristate 13-acetate ('TPA')
Ref-1	redox factor 1
RIP	receptor-interacting protein
ROS	reactive oxygen species
SAPK	stress-activated protein kinase
SAPKK	SAPK kinase
Scid mouse	severe combined immune-deficient mouse
SEK	SAPK/ERK kinase
SH3	Src homology-3
SNF	sucrose non-fermenting
SOD	superoxide dismutase
SOK	Ste20-like oxidant-stress-activated kinase
(m)SOS	(mammalian) son of sevenless
SPRK	SH3-domain-containing, proline-rich kinase
STRE	stress response element
TAK	TGF-β-activated kinase

TGF	transforming growth factor
TIPK	TNF- and IL-1-activated protein kinase
TNF	tumour necrosis factor
TNFR	TNF receptor
Tor/TOR	target of rapamycin
TRADD	TNF-receptor-associated death domain
TRAF	TNF-receptor-associated factor
TRE	TPA-responsive element
VEGF	vascular endothelial growth factor

Biochem. Soc. Symp. **64**, 1–12
Printed in Great Britain

1

Signal transduction by the c-Jun N-terminal kinase

Roger J. Davis

Howard Hughes Medical Institute and Program in Molecular Medicine, Department of Biochemistry and Molecular Biology, University of Massachusetts Medical School, 373 Plantation Street, Worcester, MA 01605, U.S.A.

Abstract

The c-Jun N-terminal kinase (JNK) group of mitogen-activated protein kinases (MAP kinases) is activated by exposure of cells to environmental stress and by the treatment of cells with cytokines. The mechanism of activation of JNK is mediated by dual phosphorylation within kinase subdomain VIII on the motif Thr-Pro-Tyr. This phosphorylation is mediated by the MAP kinase kinases MKK4 and MKK7. These MAP kinase kinases serve as signalling molecules that integrate a wide array of stimuli into the activation of the JNK signalling pathway. Studies of the physiological function of JNK have been facilitated by the molecular genetic analysis of JNK signalling in *Drosophila* and by the creation of mice with targeted disruption of components of the JNK pathway. These studies demonstrate that the JNK pathway regulates AP-1 (activator protein-1) transcriptional activity *in vivo* and indicate that JNK is required for embryonic morphogenesis, the regulation of cellular proliferation and apoptosis, and the response of cells to immunological stimuli.

Introduction

Mitogen-activated protein kinases (MAP kinases) are established to be important mediators of intracellular signalling within cells. These protein kinases function within signalling pathways that are initiated by multiple mechanisms, including the activation of cell surface receptors. A major target of MAP kinase signalling is the regulation of gene expression. These properties implicate MAP kinases in developmental processes and in the response of cells to their environment, for example growth factors, cytokines or exposure to stress. Indeed, studies using both genetic and biochemical approaches have demonstrated the essential role of MAP kinases in mammals, insects, nematodes and plants.

In mammals, three groups of MAP kinases have been identified: the extracellular-signal-regulated kinases (ERKs); the p38 MAP kinases; and the c-Jun N-terminal kinases (JNKs; also known as stress-activated protein kinases, or SAPKs) [1]. These MAP kinases are activated by dual phosphorylation within protein kinase subdomain VIII [1]. This phosphorylation is mediated by a protein kinase cascade that consists of a MAP kinase, a MAP kinase kinase and a MAP kinase kinase kinase. Individual MAP kinases are activated by different signalling modules that are regulated by different stimuli [1–3]. For example, the ERK group is activated by the MAP kinase kinases MKK1 and MKK2; the p38 MAP kinase group is activated by MKK3, MKK4 and MKK6; and the JNK group is activated by MKK4 and MKK7 (Fig. 1). These separate signalling modules allow the integrated response of MAP kinase pathways to different stimuli.

The JNK group of MAP kinases

Three genes that encode JNKs have been identified by molecular cloning. The human genes are *JNK1* [4], *JNK2* [5,6] and *JNK3* [7,8]. The corresponding genes in the rat have also been identified [9]. Transcripts of each of these genes are alternatively spliced to create mRNAs that encode 46 kDa and 55 kDa JNK isoforms. The presence of a C-terminal extension on the 55 kDa isoforms serves to distinguish these isoforms from the 46 kDa JNK isoforms [7]. An additional site of alternative splicing has

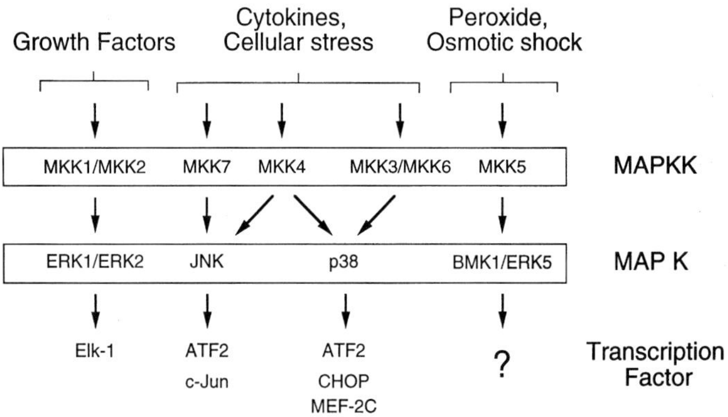

Fig. 1. Structure of mammalian MAP kinase signal transduction pathways. The ERK, BMK (big MAP kinase) 1/ERK5, p38 and JNK signal transduction pathways are illustrated schematically. MKK1 and MKK2 are activators of the ERK subgroup of MAP kinases. MKK3, MKK4 and MKK6 are activators of p38 MAP kinases. MKK5 is an activator of BMK1/ERK5. MKK7 is a specific activator of the JNK group of MAP kinases, whereas MKK4 activates both the p38 and JNK subgroups of MAP kinases. Abbreviations: CHOP, CCAAT-enhancer-binding protein homologous protein-10; MEF, myocyte-enhancing factor.

been identified within the kinase domains of JNK1 [7] and JNK2 [7,9], but not JNK3 [7]. This pattern of alternative splicing is illustrated in Fig. 2. No functional differences have been detected in experiments designed to compare the 46 kDa and 55 kDa JNK isoforms [7]. In contrast, the alternative splicing of JNK1 and JNK2 within the kinase domain causes changes in the substrate specificity of these protein kinases.

Substrate recognition by JNKs is mediated by a binding interaction between a site on the substrate and the JNK [4]. This binding site is independent of the sites of phosphorylation by JNK [3]. Deletion of the

```
JNK1-alpha                                            MSRSKRDNNFYSVEIGDSTFTVLKRYQN
JNK1-beta                                             ............................
JNK2-alpha                                            M.D..C.SQ....QVA...........Q
JNK2-beta                                             M.D..C.SQ....QVA...........Q
JNK3-alpha      MSLHFLYYCSEPTLDVKIAFCQGFDKQVDVSYIAKHYNM.K..V..Q.....V.............

                       I                   II               III               IV
JNK1-alpha      LKPIGSGAQGIVCAAYDAILERNVAIKKLSRPFQNQTHAKRAYRELVLMKCVNHKNIIGLLNVFTP
JNK1-beta       ..................................................................
JNK2-alpha      ...............F.TV.GIS..V......................L.........S.......
JNK2-beta       ...............F.TV.GIS..V......................L.........S.......
JNK3-alpha      ..................V.D.....................................S.......

                                     V               VIA                  VIB
JNK1-alpha      QKSLEEFQDVYIVMELMDANLCQVIQMELDHERMSYLLYQMLCGIKHLHSAGIIHRDLKPSNIVVK
JNK1-beta       ..................................................................
JNK2-alpha      ..T........L.............H........................................
JNK2-beta       ..T........L.............H........................................
JNK3-alpha      ..T........L......................................................

                         VII          * *   VIII              IX
JNK1-alpha      SDCTLKILDFGLARTAGTSFMMTPYVVTRYYRAPEVILGMGYKENVDLWSVGCIMGEMVCHKILFPG
JNK1-beta       ...............................................I..........IKGGV....
JNK2-alpha      ................C.N............................I.........L.KGCVI.Q.
JNK2-beta       ................C.N............................I.......A...L..V....
JNK3-alpha      ...............................................I...........R.......

                               X
JNK1-alpha      RDYIDQWNKVIEQLGTPCPEFMKKLQPTVRTYVENRPKYAGYSFEKLFPDVLFPADSEHNKLKASQ
JNK1-beta       T.H...............................................................
JNK2-alpha      T.H..............SA...........N........P.IK..E....WI..SE..RD.I.T..
JNK2-beta       .................SA...........N........P.IK..E....WI..SE..RD.I.T..
JNK3-alpha      ..............................N..........LT.P.....S...............

                               XI
JNK1-alpha      ARDLLSKMLVIDASKRISVDEALQHPYINVWYDPSEAEAPPPKIPDKQLDEREHTIEEWKELIYKE
JNK1-beta       ..................................................................
JNK2-alpha      ............PD.........R....T.....A.......Q.Y.A..E....A...........
JNK2-beta       ............PD.........R....T.....A.......Q.Y.A..E....A...........
JNK3-alpha      ............PA......DA............A.V.....Q.Y.....................

JNK1-alpha1     VMDLEERTKNGVIRGQPSPLGAAVINGSQHPSSSSSVNDVSSMSTDPTLASDTDSSLEAAAGPLGCCR
JNK1-alpha2     ....................AQVQQ
JNK1-beta1      ....................................................................
JNK1-beta2      ....................AQVQQ
JNK2-alpha1     ...W...S....VKD...D--...SSNA-T..Q...I..I.....EQ..........DAST...EG..
JNK2-alpha2     ...W...S....VKD...--ALPHAQMQQ
JNK2-beta1      ...W...S....VKD...D--...SSNA-T..Q...I..I.....EQ..........DAST...EG..
JNK2-beta2      ...W...S....VKD...--ALPHAQMQQ
JNK3-alpha1     ..NS..K.....VK.....S....NSSE-SL.P......I......Q............S........
JNK3-alpha2     ..NS..K.....VK.....SAQVQQ
```

Fig. 2. Comparison of the primary structures of JNKs. The primary structures of human JNK1, JNK2 and JNK3 are compared using the PILE-UP program (version 7.2; Wisconsin Genetics Computer Group). Gaps introduced into the sequences to optimize the alignment are illustrated with a dash (–). The sites of activating phosphorylation of JNK [4] are indicated with asterisks (*).

JNK binding site prevents phosphorylation of the substrate by JNK [3]. The alternative splicing of JNK1 and JNK2 within the kinase domain changes the specificity of the binding interaction between JNK and its substrates [5,7,10]. This observation suggests that indivivdual JNK isoforms target different groups of JNK substrates *in vivo*.

The JNK MAP kinases are activated by exposure of cells to environmental stress or by treatment of cells with pro-inflammatory cytokines [4–7,9]. Targets of the JNK signal transduction pathway include the transcription factors ATF2 (activating transcription factor 2) and c-Jun [1]. These transcription factors are members of the bZIP (basic region leucine zipper) group that bind as homo- and hetero-dimeric complexes to AP-1 and AP-1-like sites in the promoters of many genes [11]. JNK binds to an N-terminal region of ATF2 and c-Jun, and phosphorylates two sites within the activation domain of each transcription factor [4,12–14]. This phosphorylation leads to increased transcriptional activity [1]. AP-1 transcriptional activity is also increased by the JNK pathway through increased expression of c-Fos and c-Jun [1]. An increase in c-Fos expression is mediated by activation of the serum response element in the c-Fos promoter [15–17]. Increased expression of c-Jun is mediated by at least two mechanisms. First, JNK causes increased AP-1 activity, which increases c-Jun expression through the AP-1-like sites in the c-Jun promoter [1]. Secondly, the phosphorylation of c-Jun by JNK causes decreased ubiquitin-mediated degradation of c-Jun and an increase in the half-life of the c-Jun protein [18,19].

Together, these biochemical studies indicate that the JNK signal transduction pathway contributes to the regulation of AP-1 transcriptional activity in response to cytokines and environmental stress [1]. Strong support for this hypothesis is provided by genetic evidence indicating that the JNK signalling pathway is required for the normal regulation of AP-1 transcriptional activity [20].

MKK4 is an activator of the JNK and p38 MAP kinases

JNK is activated by dual phosphorylation on Thr-183 and Tyr-185 [4]. This phosphorylation is mediated, in part, by the MAP kinase kinase MKK4 (also known as SEK1) [2,21,22]. The relationship between the primary sequences of MKK4 and other members of the MAP kinase kinase group is illustrated in Fig. 3. MKK4 phosphorylates and activates JNK *in vitro*. Furthermore, transfection studies demonstrate that MKK4 activates JNK *in vivo*. In addition, transfection of cells with dominant-negative MKK4 inhibits JNK activation *in vivo*. These studies provide biochemical evidence for a role for MKK4 as an activator of JNK [1].

In vitro protein kinase assays demonstrate that MKK4 is also able to activate the p38 MAP kinases [2,21]. Transfection assays indicate that MKK4 can act as an activator of the p38 MAP kinase pathway. MKK4 is therefore a candidate activator of the p38 MAP kinase pathway *in vivo*.

```
              1
MKK7                                                                MLGLPSTLFTPRSMES
HEP           <SASSSSSSASAFASAAPATGTFGGTYTPPTTRVSRATPTLPMLSSGPGGGLNRTRPVILP.P..PHPPV
MKK1                                                     MPKKKP--TPIQ.NPA-PDGSAVNG
MKK2                                                    MLARRKPVLPA.TINP.IAEGP.PT.
MKK3                                                             MSKPPA-----PN.TPPRN
MKK4          MAAPSPSGGGGSGGGSGSGTPGPVGSPAPGHPAVSSMQGKRKALKLNFANPPFKSTARFTLNPN.TGVQN
MKK5          <IGQVLPEATTTAFEYEDEDGDRITVRSDEEMKAMLSYYYSTVMEQQVNGQLIEP.QIFPRACK.PGERN
MKK6                                               MSQSKGKKRNPGLKIPKEAFEQPQ-----TSSTPPRD
Consensus

              71                                                         I
MKK7          IEIDQKLQEIMKQT-GYLTIG-----GQRYQAEI-------------NDLENLGEMGSGTCGQVWKMRFR
HEP           S.T.M..KI..E..-.K.N.N-----.RQ.PTD.-------------...KH..DL.N..S.N.V..MHL
MKK1          TSSAETNL.ALQKKLEE.ELDE----Q..KRL.AFLTQKQKVGELKDD.F.KIS.L.A.NG.V.F.VSHK
MKK2          EGASEANLVDLQKKLEE.ELDE----Q.KKRL.AFLTQKAKVGELKDD.F.RIS.L.A.NG.V.T.VQH.
MKK3          ----------LDSR-TFI..G-----DRNFEV.A-------------D..VTIS.L.R.AY.V.E.V.HA
MKK4          PH.ERLRTHSIESS-.K.K.SP----E.HWDFTA-------------E..KD...I.R.AY.S.N..VHK
MKK5          .HGLKVNTRAGPSQHSSPAVSDSLPSNSLKKSSAELKKILANGQMNEQ.IRYRDTL.H.NG.T.Y.AYHV
MKK6          ----------LDSK-ACIS.G-----N.NFEVKA-------------D.L.PIM.L.R.AY.V.E...HV
Consensus                                                     D       G G  G V K

              141   II                 III           IV                V
MKK7          KTGHIIAVKQMRRSGNKEENKRILMDLDVVLKSHDCPYIVQCFGTFITNTDVFIAMELM-GTCAEKLKK-
HEP           SSNT........T..A.................K...K.L.C.VRDP..W.C....-SM.FD..L.-
MKK1          PS.LVM.R.LIHLEIKPAIRNQ.IRE.Q.-.HECNSP...GFY.A.YSDGEIS.C..H.D.GSLDQVL.K
MKK2          PS.L.M.R.LIHLEIKPAIRNQ.IRE.Q.-.HECNSP...GFY.A.YSDGEIS.C..H.D.GSLDQVL.E
MKK3          QS.T.M...RI.A.V.SQ.Q..LL....INMRTV..F.T.TFY.ALFREG..W.C....D-.SLD.FYRK
MKK4          PS.Q.M...RI.S.VDEK.Q.QLL......MRSS.......FY.ALFREG.CW.C....-S.SFD.FY.Y
MKK5          PS.K.L...VILLDITL.LQ.Q.MSE.EILI.-C.SS..IGFY.A.FVENRIS.CTEF.D.GSLDDIG.-
MKK6          PS.Q.M...RI.A.V.SQ.Q..LL....ISMRTV...FTVTFY.ALFREG..W.C....-D-.SLD.FY.Q
Consensus         I A K                 L                G         I    M

              211                               VI                  VII        *   *
MKK7          ---RMQGPIPERILGKMTVAIVKALYYLKEKHGVIHRDVKPSNILLDERGQIKLCDFGISGRLVDSKAKT
HEP           ---LSKK.V..Q....V...T.N..S...D...............I....N.................N.
MKK1          -----A.R...Q....VSI.VIKG.T..R...KIM..........VNS..E.......V..Q.I..M.NS
MKK2          -----AKR...E....VSI.VLRG.A..R...QIM..........VNS..E.......V..Q.I..M.NS
MKK3          VLDK-NMT...D...EIA.S..R..EH.HS.LS.I........V.INKE.HV.M.......Y....V...
MKK4          VYSVLDDV...E....I.L.T.K..NH...NLKII...I.......DRS.N..........Q....I...
MKK5          --------M..HV..RIA..V.KG.T..WS-LKIL.......M.VNT...V......V.TQ..N.I...
MKK6          VIDK-GQT...D....IA.S.....EH.HS.LS.I........V.INAL..V.M.......Y....V...
Consensus              PE ILG         L  L       HRD KPSN L    G  K CDFG S  LV S A

              281      VIII                IX
MKK7          RSAGCAAYMAPERIDPPDPTKPDYDIRADVWSLGISLVELATGQFPYKNCKTD-----------------
HEP           ..............---..K..K............T......ARS..EG.N..-----------------
MKK1          F-V.TRS..S...LQGTH-----.SVQS.I..M.L....M.V.RY.IPPPDAKELELMFGCQV----EGD
MKK2          F-V.TRS......LQGTH-----.SVQS.I..M.L......V.RY.IPPPDAKELEAIFGRPVVDGEEGE
MKK3          MD...KP.......N-.ELNQKG.NVKS.......TMI.M.ILR...ESWG.P-----------------
MKK4          .D...RP........-.SASRQG..V.S.......T.Y.....R...PKWNSV-----------------
MKK5          Y-V.TN........SGEQ-----.G.HS........FM.IQKN.GSLMP---------------------
MKK6          ID...KP.......N-.ELNQKG.SVKS.I.....TMI...ILR...DSWG.P-----------------
Consensus        G   YM PER          Y    D WSLG    E

              351                            X                                   XI
MKK7          ----------------------------FEVLTKVLQEEPPLLP-G-HMGFSGDFQSFVKDCLTKDHRKR
HEP           ----------------------------........DS...C..Y.EGYN..QQ.RD..IK....N.QD.
MKK1          AAETPPRPRTPGRPLSSYGMDSRPPMAI..L.DYIVN.P..K..S.V---..LE..D..NK..I.NPAE.
MKK2          PHSISPRPRPPGRPVSGHGMDSRPAMAI..L.DYIVN.P..K..N.V---.TP...E..NK..I.NPAE.
MKK3          ----------------------------.QQ.KQ.VE.PS.Q..AD---R..PE.VD.TAQ..R.NPAE.
MKK4          ----------------------------.DQ..Q.VKGD..Q.SNSEERE..PS.IN..NL.....ES..
MKK5          ----------------------------LQL.QCIVD.DS.V..V.E---..EP.VH.ITQ.MR.QPKE.
MKK6          ----------------------------.QQ.KQ.V...S.Q..AD---K..AE.VD.TSQ..K.NSKE.
Consensus                                    L        P L       F      F   CL K    R

              421
MKK7          PKYNKLLEHSFIKHYEILEVDVAS-WFKDVMAKTESPRTSGVLSQHHLPFFR
HEP           ...PE..AQP..RI..SAK...PN-..QSI--.DNRL.AN.DPTLQR..NS
MKK1          ADLKQ.MV.A...RSDAE...F.G-.LCSTIGLNQPSTPTHAAGV
MKK2          ADLKM.TN.T...RS.VE...F.G-.LCKTLRLNQPGTPTRTAV
MKK3          MS.LE.M..P.FTLHKTKKT.I.A-FV.KILGEDS
MKK4          ...KE..K.P..LMY.ERA.E..C-YVCKILDQMPATP.SPMYVD
MKK5          .APEE.MG.P..VQFNDGNAA.VSM.VCRALEERRTSRGPREAAAGH
MKK6          .T.PE.MQ.P.FTLH.SKGT....-FVKLILGD
Consensus          L    F
```

Fig. 3. Comparison of the primary structures of MAP kinase kinases. The mammalian MAP kinase kinase group includes at least seven members. The first group of MAP kinases that was identified comprised MKK1 and MKK2 [75]. These protein kinases function as activators of the ERK group of MAP kinases. Subsequent studies demonstrated that the p38 MAP kinases are activated by MKK3, MKK4 and MKK6; the BMK1/ERK5 MAP kinase is activated by MKK5; and the JNK group of MAP kinases is activated by MKK4 and MKK7. Each of these MAP kinase kinases functions as an activator of a single group of MAP kinases, with the exception of MKK4, which phosphorylates and activates both JNK and p38 MAP kinases. The figure presents a comparison of the primary structure of MKK7 with those of the MAP kinase kinases HEP (hemipterous), MKK1, MKK2, MKK3, MKK4, MKK5 and MKK6 using the PILE-UP program (version 7.2; Wisconsin Genetics Computer Group). Gaps introduced into the sequences to optimize alignment are illustrated with a dash (—). The sites of activating phosphorylation of MAP kinase kinases [2,25,76,77] are indicated with asterisks (*).

However, two lines of evidence suggest that the role of MKK4 as a regulator of the p38 MAP kinase pathway requires further evaluation. First, dominant-negative MKK4 inhibits JNK activity more potently than it inhibits p38 MAP kinase. This differential inhibition of JNK and p38 MAP kinases may be mediated by the stronger binding interaction that is observed between MKK4 and JNK than between MKK4 and p38 MAP kinase. Secondly, an upstream activator of MKK4, MEKK1 (MEK kinase 1) [23,24], causes selective activation of the JNK signal transduction pathway in transfection experiments [2,21]. Together, these data indicate that, while MKK4 does activate p38 MAP kinase *in vitro*, the role of MKK4 in the regulation of p38 MAP kinase *in vivo* is unclear. Other studies have implicated the MAP kinase kinases MKK3 [2] and MKK6 [25–28] as major activators of the p38 MAP kinase *in vivo*.

Recently, two studies using homologous recombination to create null alleles of the *MKK4* gene in murine cells have been reported [20,29]. Disruption of the *MKK4* gene caused embryonic death in mice [20]. Analysis of cells with a homozygous deficiency in MKK4 demonstrated that MKK4 is required for the normal regulation of JNK by environmental stress [20,29]. These data provide strong genetic evidence that MKK4 does function as an activator of the JNK signal transduction pathway *in vivo*. No defects in the regulation of the p38 MAP kinase signalling pathway were detected in MKK4 (−/−) cells [20]. The normal regulation of p38 MAP kinase activity in MKK4 (−/−) cells may be caused either: (1) by the absence of a role for MKK4 in p38 MAP kinase regulation *in vivo*; or (2) by complementation of the MKK4 defect by MKK3 and MKK6. Further studies are required to discriminate between these two possible mechanisms.

MKK7 is an activator of the JNK signal transduction pathway

The existence of a specific activator of JNK that is independent of MKK4 has been proposed previously [2]. Indeed, biochemical studies support the conclusion that MKK4 is not the only activator of JNK in mammalian cells [30,31]. Furthermore, genetic evidence for this novel JNK activator was obtained from the results of experiments in which the *MKK4* gene was disrupted by homologous recombination [20,29]. These studies demonstrated that, although MKK4 (−/−) cells are defective in JNK regulation, the loss of MKK4 does not block JNK activation completely [20,29]. They provide definitive evidence that MKK4 represents only one mechanism of activation of the JNK protein kinase *in vivo*.

Recently, the molecular cloning of a new member of the mammalian MAP kinase kinase group, MKK7, was reported [32]. Comparison of MKK7 with other members of the mammalian MAP kinase kinase group indicates that it is related to the JNK activator MKK4 (Figs. 3 and 4). However, MKK7 is most closely related to the *Drosophila* protein kinase hemipterous (HEP) [32,33].

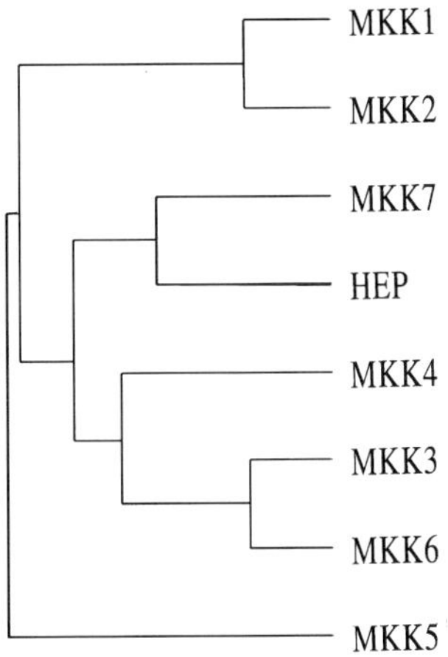

Fig. 4. Relationship between members of the MAP kinase kinase group. The dendrogram was created by the unweighted pair-group method using arithmetic averages (PILE-UP program; version 7.2; Wisconsin Genetics Computer Group). The MAP kinase kinases MKK1, MKK2, MKK3, MKK4, MKK5, MKK6 and MKK7 and the *Drosophila* MAP kinase kinase HEP are presented.

The MAP kinase kinase MKK7 activates JNK, but not the ERK or p38 MAP kinases, *in vitro*. Transfection assays confirm that MKK7 activates JNK, but not the ERK or p38 MAP kinases, *in vivo*. MKK7 therefore appears to be the specific JNK activator that has been proposed in previous studies [2]. Thus mammalian cells express two activators of the JNK MAP kinases, MKK4 and MKK7 (Fig. 1).

MKK4 and MKK7 integrate signals initiated at the cell surface

MKK4 and MKK7 function as activators of JNK. Northern blot analysis demonstrates that both MKK4 and MKK7 are widely expressed in human and murine tissues [32]. Although MKK4 and MKK7 are co-expressed, the relative abundance of each MAP kinase kinase differs between tissues [32]. These data indicate that MKK4 and MKK7 serve to integrate the divergent signals that lead to activation of the JNKs *in vivo*. Further studies are required to identify the signalling mechanisms that lead to the activation of MKK4, MKK7, and MKK4 together with MKK7.

Multiple protein kinases have been reported to function as upstream elements of the JNK signalling pathway (reviewed in [1,34]). The mixed-lineage kinases MLK2 [35], MLK3 [36,37] and DLK/MUK (dual leucine-zipper kinase) [38] function as activators of the JNK pathway that phosphorylate and activate MKK4. The protein kinases ASK1 (apoptosis-stimulating kinase 1) [39] and Tpl-2/Cot [40] also activate the MKK4/JNK pathway. In addition, members of the MEKK family [41], including MEKK1 [23,24], MEKK2 [42,43], MEKK3 [42,43] and MEKK4 [44], activate the JNK signalling pathway by phosphorylation and activation of MKK4. The primary role for MKK4 inferred in each of these studies will need to be re-evaluated in light of the recent identification of the second JNK activator, MKK7 [32].

Other protein kinases have also been implicated as activators of the JNK pathway, although the mechanism by which they activate JNK is unclear. Examples of these kinases include members of the large group of Ste20-related protein kinases, e.g. PAK (p21-activated kinase) [45,46], GCK (germinal centre kinase) [47], Krs (kinase responsive to stress) [48] and NIK (Nck-interacting kinase) [49]. One additional Ste20-related protein kinase that activates the JNK pathway, HPK1 (haematopoietic progenitor kinase 1), appears to activate JNK through interaction with the mixed-lineage kinase MLK3 [50–52].

The mechanism by which environmental stress causes activation of the JNK signalling pathway is unclear, but it has been proposed that this response is mediated by the clustering of cell surface receptors [53]. The signalling mechanisms employed by cell surface receptors to activate the JNK pathway are poorly understood. However, recent studies have led to the elucidation of receptor-proximal components that lead to the activation of the JNK signalling pathway. For example, the adaptor protein TRAF2 (tumour necrosis factor-receptor-associated factor) [54–56] and the protein kinase RIP (receptor-interacting protein) [54] couple tumour necrosis factor receptor signalling to the JNK pathway. The effect of TRAF2 appears to be mediated by a MAP kinase kinase kinase, which binds to TRAF2 [57]. In the case of Fas signalling, the protein Daxx binds to the death domain of Fas and mediates activation of the JNK pathway [58]. Other upstream components of the JNK signalling pathway include the small GTPases Rac1 and Cdc42 [45,46,59,60] and heterotrimeric GTPases [61–63].

Function of the JNK signal transduction pathway

Comparison of the primary sequence of MKK7 with those of other members of the mammalian MAP kinase kinase group demonstrates that MKK7 is most similar to MKK4 (Fig. 1). This similarity in primary sequence reflects the common enzymic property of MKK4 and MKK7 as activators of JNK. However, MKK7 is most similar to the *Drosophila* MAP kinase kinase HEP. Biochemical analysis of HEP demonstrates that it is a potent activator of *Drosophila* JNK *in vitro* [64]. The function of HEP as a physiological JNK activator is supported by genetic analysis. Loss-of-function alleles of the MAP kinase kinase (*hep*) and JNK (*bsk*) cause the same embryonic-lethal phenotype [33,64,65]. Detailed studies of *Drosophila* development demonstrate that JNK is required for morphogenetic cell movement during embryogenesis [33,64,65]. Similarly, the JNK pathway is required for embryonic viability in mice [20]. It is therefore possible that, as in *Drosophila*, the JNK signalling pathway is required for embryonic morphogenesis in mammals. The specific targets of the JNK signalling pathway during embryonic development remain to be identified.

The function of the JNK signal transduction pathway *in vivo* is poorly understood. Gene disruption experiments in mice demonstrate that the JNK pathway is required for the normal regulation of AP-1 transcriptional activity [20]. The biological significance of AP-1 regulation by the JNK pathway is unclear, but this pathway has been implicated in the stress-induced apoptosis of neurons [66] and other cell types [67,68], transformation of pre-B-cells by *Bcr–Abl* [69,70], transformation of fibroblasts by the *Met* oncogene [71], survival signals in cells exposed to apoptotic stimuli [1,29,54], co-stimulatory signalling in the immune response of T-cells [72,73], and the inflammation-associated expression of E-selectin by endothelial cells [74]. Further studies of the physiological role of the JNK signal transduction pathway in mammals will be greatly facilitated by the creation of animals with specific defects in JNK signalling. Targeted disruption of the *MKK4* gene causes only a partial defect in JNK signalling [20,29]. Progress towards understanding the physiological role of the JNK signalling pathway will require analysis of the effects of targeted disruption of the genes encoding MKK7, JNK1, JNK2 and JNK3.

I thank K. Gemme for administrative assistance. The studies performed in this laboratory were supported by grants from the National Cancer Institute. R.J.D. is an investigator of the Howard Hughes Medical Institute.

References

1. Whitmarsh, A.J. and Davis, R.J. (1996) J. Mol. Med. **74**, 589–607
2. Dérijard, B., Raingeaud, J., Barrett, T., Wu, I.-H., Han, J., Ulevitch, R.J. and Davis, R.J. (1995) Science **267**, 682–685
3. Davis, R.J. (1994) Trends Biochem. Sci. **19**, 470–473
4. Dérijard, B., Hibi, M., Wu, I.-H., Barrett, T., Su, B., Deng, T., Karin, M. and Davis, R.J. (1994) Cell **76**, 1025–1037
5. Kallunki, T., Su, B., Tsigelny, I., Sluss, H.K., Dérijard, B., Moore, G., Davis, R.J. and Karin, M. (1994) Genes Dev. **8**, 2996–3007
6. Sluss, H.K., Barrett, T., Dérijard, B. and Davis, R.J. (1994) Mol. Cell. Biol. **14**, 8376–8384
7. Gupta, S., Barrett, T., Whitmarsh, A.J., Cavanagh, J., Sluss, H.K., Dérijard, B. and Davis, R.J. (1996) EMBO J. **15**, 2760–2770
8. Mohit, A.A., Martin, J.H. and Miller, C.A. (1995) Neuron **14**, 67–78
9. Kyriakis, J.M., Banerjee, P., Nikolakaki, E., Dai, T., Rubie, E.A., Ahmad, M.F., Avruch, J. and Woodgett, J.R. (1994) Nature (London) **369**, 156–160
10. Dai, T., Rubie, E., Franklin, C.C., Kraft, A., Gillespie, D.A., Avruch, J., Kyriakis, J.M. and Woodgett, J.R. (1995) Oncogene **10**, 849–855
11. Curran, T. and Franza, B.J. (1988) Cell **55**, 395–397
12. van Dam, H., Wilhelm, D., Herr, I., Steffen, A., Herrlich, P. and Angel, P. (1995) EMBO J. **14**, 1798–1811
13. Gupta, S., Campbell, D., Dérijard, B. and Davis, R.J. (1995) Science **267**, 389–393
14. Livingstone, C., Patel, G. and Jones, N. (1995) EMBO J. **14**, 1785–1797

15. Cavigelli, M., Dolfi, F., Claret, F.-X. and Karin, M. (1995) EMBO J. **14**, 5957–5964
16. Whitmarsh, A.J., Shore, P., Sharrocks, A.D. and Davis, R.J. (1995) Science **269**, 403–407
17. Gille, H., Strahl, T. and Shaw, P.E. (1995) Curr. Biol. **5**, 1191–1200
18. Musti, A.M., Treier, M. and Bohmann, D. (1997) Science **275**, 400–402
19. Fuchs, S.Y., Dolan, L., Davis, R.J. and Ronai, Z. (1996) Oncogene **13**, 1531–1535
20. Yang, D., Tournier, C., Wysk, M., Lu, H.-T., Xu, J., Davis, R.J. and Flavell, R.A. (1997) Proc. Natl. Acad. Sci. U.S.A. **94**, 3004–3009
21. Lin, A., Minden, A., Martinetto, H., Claret, F.-X., Lange-Carter, C., Mercurio, F., Johnson, G.L. and Karin, M. (1995) Science **268**, 286–290
22. Sanchez, I., Hughes, R.T., Mayer, B.J., Yee, K., Woodgett, J.R., Avruch, J., Kyriakis, J.M. and Zon, L.I. (1994) Nature (London) **372**, 794–798
23. Yan, M., Dai, T., Deak, J.C., Kyriakis, J.M., Zon, L.I., Woodgett, J.R. and Templeton, D.J. (1994) Nature (London) **372**, 798–800
24. Minden, A., Lin, A., McMahon, M., Lange-Carter, C., Dérijard, B., Davis, R.J., Johnson, G.L. and Karin, M. (1994) Science **266**, 1719–1723
25. Raingeaud, J., Whitmarsh, A.J., Barrett, T., Dérijard, B. and Davis, R.J. (1996) Mol. Cell. Biol. **16**, 1247–1255
26. Han, J., Lee, J.-D., Jiang, Y., Li, Z., Feng, L. and Ulevitch, R.J. (1996) J. Biol. Chem. **271**, 2886–2891
27. Stein, B., Brady, H., Yang, M.X., Young, D.B. and Barbosa, M.S. (1996) J. Biol. Chem. **271**, 11427–11433
28. Moriguchi, T., Kuroyanagi, N., Yamaguchi, K., Gotoh, Y., Irie, K., Kano, T., Shirakabe, K., Muro, Y., Shibuya, H., Matsumoto, K., et al. (1996) J. Biol. Chem. **271**, 13675–13679
29. Nishina, H., Fischer, K.D., Radvanyl, L., Shahinian, A., Hakem, R., Ruble, E.A., Bernstein, A., Mak, T.W., Woodgett, J.R. and Penninger, J.M. (1997) Nature (London) **385**, 350–353
30. Meier, R., Rouse, J., Cuenda, A., Nebreda, A.R. and Cohen, P. (1996) Eur. J. Biochem. **236**, 769–805
31. Moriguchi, T., Kawasaki, H., Matsuda, S., Gotoh, Y. and Nishida, E. (1995) J. Biol. Chem. **270**, 12969–12972
32. Tournier, C., Whitmarsh, A.J., Cavanagh, J., Barrett, T. and Davis, R.J. (1997) Proc. Natl. Acad. Sci. U.S.A. **94**, 7337–7342
33. Glise, B., Bourbon, H. and Noselli, S. (1995) Cell **83**, 451–461
34. Fanger, G.R., Gerwins, P., Widmann, C., Jarpe, M.B. and Johnson, G.L. (1997) Curr. Opin. Genet. Dev. **7**, 67–74
35. Hirai, S., Katoh, M., Terada, M., Kyriakis, J.M., Zon, L.I., Rana, A., Avruch, J. and Ohno, S. (1997) J. Biol. Chem. **272**, 15167–15173
36. Rana, A., Gallo, K., Godowski, P., Hirai, S.-I., Zon, L., Kyriakis, J.M. and Avruch, J. (1996) J. Biol. Chem. **271**, 19025–19028
37. Teramoto, H., Coso, O.A., Miyata, H., Igishi, T., Miki, T. and Gutkind, J.S. (1996) J. Biol. Chem. **271**, 27225–27228
38. Hirai, S., Izawa, M., Osada, S., Spyrou, G. and Ohno, S. (1996) Oncogene **12**, 641–650
39. Ichijo, H., Nishida, E., Irie, K., ten Dijke, P., Saitoh, M., Moriguchi, T., Takagi, M., Matsumoto, K., Miyazono, K. and Gotoh, Y. (1997) Science **275**, 90–94
40. Salmeron, A., Ahmad, T.B., Carlile, G.W., Pappin, D., Narsimhan, R.P. and Ley, S.C. (1996) EMBO J. **15**, 817–826
41. Lange-Carter, C.A., Pleiman, C.M., Gardner, A.M., Blumer, K.J. and Johnson, G.L. (1993) Science **260**, 315–319

42. Deacon, K. and Blank, J.L. (1997) J. Biol. Chem. **272**, 14489–14496
43. Blank, J.L., Gerwins, P., Elliott, E.M., Sather, S. and Johnson, G.L. (1996) J. Biol. Chem. **271**, 5361–5368
44. Gerwins, P., Blank, J.L. and Johnson, G.L. (1997) J. Biol. Chem. **272**, 8288–8295
45. Zhang, S., Han, J., Sells, M.-A., Chernoff, J., Knaus, U.G., Ulevitch, R.J. and Bokoch, G.M. (1995) J. Biol. Chem. **270**, 23934–23936
46. Bagrodia, S., Derijard, B., Davis, R.J. and Cerione, R.A. (1995) J. Biol. Chem. **270**, 27995–27998
47. Pombo, C.M., Kehrl, J.H., Sanchez, I., Katz, P., Avruch, J., Zon, L.I., Woodgett, J.R., Force, T. and Kyriakis, J.M. (1995) Nature (London) **377**, 750–754
48. Tung, R.M. and Blenis, J. (1997) Oncogene **14**, 653–659
49. Su, Y.C., Han, J., Xu, S., Cobb, M. and Skolnik, E.Y. (1997) EMBO J. **16**, 1279–1290
50. Hu, M.C., Qiu, W.R., Wang, X., Meyer, C.F. and Tan, T.H. (1996) Genes Dev. **10**, 2251–2264
51. Kiefer, F., Tibbles, L.A., Anafi, M., Janssen, A., Zanke, B.W., Lassam, N., Pawson, T., Woodgett, J.R. and Iscove, N.N. (1996) EMBO J. **15**, 7013–7025
52. Tibbles, L.A., Ing, Y.L., Kiefer, F., Chan, J., Iscove, N., Woodgett, J.R. and Lassam, N.J. (1996) EMBO J. **15**, 7026–7035
53. Rosette, C. and Karin, M. (1996) Science **274**, 1194–1197
54. Liu, Z.-G., Hsu, H., Goeddel, D.V. and Karin, M. (1996) Cell **87**, 565–576
55. Reinhard, C., Shamoon, B., Shyamala, V. and Williams, L.T. (1997) EMBO J. **16**, 1080–1092
56. Natoli, G., Costanzo, A., Ianni, A., Templeton, D.J., Woodgett, J.R., Balsano, C. and Levrero, M. (1997) Science **275**, 200–203
57. Malinin, N.L., Boldin, M.P., Kovalenko, A.V. and Wallach, D. (1997) Nature (London) **385**, 540–544
58. Yang, X., Khosravi-Far, R., Chang, H.Y. and Baltimore, D. (1997) Cell **89**, 1067–1076
59. Coso, O.A., Chiariello, M., Yu, J.-C., Teramoto, H., Crespo, P., Xu, N., Miki, T. and Gutkind, J.S. (1995) Cell **81**, 1137–1146
60. Minden, A., Lin, A., Claret, F.-X., Abo, A. and Karin, M. (1995) Cell **81**, 1147–1157
61. Shapiro, P.S., Evans, J.N., Davis, R.J. and Posada, J.A. (1996) J. Biol. Chem. **271**, 5750–5754
62. Coso, O.A., Teramoto, H., Simonds, W.F. and Gutkind, J.S. (1996) J. Biol. Chem. **271**, 3963–3966
63. Prasad, M.V., Dermott, J.M., Heasley, L.E., Johnson, G.L. and Dhanasekaran, N. (1995) J. Biol. Chem. **270**, 18655–18659
64. Sluss, H.K., Han, Z., Barrett, T., Davis, R.J. and Ip, T. (1996) Genes Dev. **10**, 2745–2758
65. Riesgo-Escovar, J.R., Jenni, M., Fritz, A. and Hafen, E. (1996) Genes Dev. **10**, 2759–2768
66. Xia, Z., Dickens, M., Raingeaud, J., Davis, R.J. and Greenberg, M.E. (1995) Science **270**, 1326–1331
67. Zanke, B.W., Boudreau, K., Rubie, E., Winnett, E., Tibbles, L.A., Zon, L., Kyriakis, J., Liu, F.-F. and Woodgett, J.R. (1996) Curr. Biol. **6**, 606–613
68. Verheij, M., Bose, R., Lin, X.H., Yao, B., Jarvis, D., Grant, S., Birrer, M.J., Szabo, E., Zon, L.I., Kyriakis, J.M., et al. (1996) Nature (London) **380**, 75–79
69. Raitano, A.B., Halpern, J.R., Hambuch, T.M. and Sawyers, C.L. (1995) Proc. Natl. Acad. Sci. U.S.A. **92**, 11746–11750
70. Dickens, M., Rogers, J.S., Cavanagh, J., Raitano, A., Xia, Z., Halpern, J.R., Greenberg, M., Sawyers, C.L. and Davis, R.J. (1997) Science **277**, 693–696

71. Rodrigues, G.A., Park, M. and Schlessinger, J. (1997) EMBO J. **16**, 2634–2645
72. Su, B., Jacinto, E., Hibi, M., Kallunki, T., Karin, M. and Ben-Neriah, Y. (1994) Cell **77**, 727–736
73. Rincon, M., Dérijard, B., Chow, C.W., Davis, R.J. and Flavell, R.A. (1997) Genes Funct. **1**, 51–68
74. Read, M.A., Whitely, M.Z., Gupta, S., Pierce, J.W., Best, J., Davis, R.J. and Collins, T. (1997) J. Biol. Chem. **272**, 2753–2761
75. Ahn, N.G., Seger, R. and Krebs, E.G. (1992) Curr. Opin. Cell Biol. **4**, 992–999
76. Alessi, D.R., Saito, Y., Campbell, D.G., Cohen, P., Sithanandam, G., Rapp, U., Ashworth, A., Marshall, C.J. and Cowley, S. (1994) EMBO J. **13**, 1610–1619
77. Zheng, C.F. and Guan, K.L. (1994) EMBO J. **13**, 1123–1131

Biochem. Soc. Symp. **64**, 13–27
Printed in Great Britain

2

Roles of the AMP-activated/SNF1 protein kinase family in the response to cellular stress

D. Grahame Hardie

Biochemistry Department, University of Dundee, Dundee DD1 4HN, Scotland, U.K.

Abstract

The AMP-activated protein kinase (AMPK) in mammals, and its homologue in *Saccharomyces cerevisiae*, are activated by cellular stresses associated with ATP depletion. AMPK is a heterotrimer comprising a catalytic α subunit with associated β and γ subunits, these being homologous with the products of the *SNF1*, *SIP1/SIP2/GAL83* and *SNF4* genes in *S. cerevisiae*. The α subunit has at least two isoforms (α1 and α2), which differ in their AMP-dependence and subcellular localization, with α2 complexes being partly nuclear. AMPK is activated allosterically by 5′-AMP, which also promotes phosphorylation and activation by an upstream kinase, and inhibits dephosphorylation and inactivation. Elevation of AMP always accompanies depletion of ATP due to the action of adenylate kinase. Since high ATP antagonizes the activating effects of AMP, the system behaves like a cellular 'fuel gauge'. It is activated by various types of stress associated with ATP depletion, such as hypoxia, heat shock, metabolic poisoning and, in muscle, exercise. AMPK phosphorylates multiple targets which switch off anabolic pathways and switch on alternative catabolic pathways. The yeast SNF1 complex is switched on by glucose starvation, and its targets include transcription factors that repress transcription of genes required for catabolism of alternative carbon sources.

Introduction

Although it is now well known that several protein kinase cascades are activated in response to different types of cellular stress [1], in most cases the intracellular signals that activate the cascades remain unknown. An exception to this is the mammalian AMP-activated protein kinase

(AMPK) system [2,3], which is activated in response to stresses that deplete cellular energy charge via mechanisms that are now well understood. AMPK was discovered by its ability to regulate metabolic enzymes involved in lipid metabolism, and these remain its best characterized physiological targets [4]. However, there are increasing indications, discussed below, that it also regulates gene expression, probably by direct phosphorylation of transcription factors. The evidence for this is particularly strong in the case of the budding-yeast homologue of AMPK, i.e. the SNF1 complex. In this article I will review current ideas concerning the structure, regulation and physiological roles of AMPK in mammals and of the SNF1 complex in yeast.

Structure of mammalian AMPK

Although some of the biochemical activities of AMPK were reported as long ago as 1973 [5,6], the kinase was not purified to homogeneity until 1994 [7,8]. It was found to consist of three subunits, i.e. α (63 kDa), β (38 kDa) and γ (35 kDa) (the molecular masses given are those estimated by SDS/PAGE; the mass of β predicted from the DNA sequence is only 30 kDa [9,10]). Based on Coomassie Blue staining, the α-, β- and γ-subunits are present in a 1:1:1 molar ratio. The native molecular mass of the holoenzyme was estimated by gel filtration and glycerol gradient centrifugation to be 190 000 Da, indicating that the holoenzyme is most likely to be an $\alpha\beta\gamma$ heterotrimer. The α-subunit is the catalytic subunit, containing a kinase domain at the N-terminus, followed by a C-terminal domain of approximately equal size [11,12]. The C-terminal domain makes interactions with β/γ [13], but other than this its function in mammals is not well understood. The functions of the β- and γ-subunits are also not understood, except that both appear to be essential for kinase activity. Using expression in CCL13 cells, Woods et al. [14] found that significant levels of neither activity nor protein were recovered if α was expressed either on its own or together with β or γ. Only when all three subunits were co-expressed were significant levels of protein and activity recovered [9]. These results suggest that the β- and γ-subunits stabilize, as well as activate, the α-subunit. Dyck et al. [13] reported similar results in COS7 cells, except that in this case detectable levels of activity were obtained when α was expressed with β or γ alone. Nevertheless, the activity was 50–100-fold higher when α was co-expressed with both β and γ. The two groups therefore agree that the β- and γ-subunits together cause a very large increase in activity over that seen with α, $\alpha\beta$ or $\alpha\gamma$ alone.

As might have been anticipated, the α-, β- and γ-subunits of mammalian AMPK were subsequently found to exist as distinct isoforms encoded by different genes. AMPK-α1 (encoded in humans on chromosome 5, p11–p14 [15]) and α2 (encoded at 1p31 [16]) are 91% identical in amino acid sequence within the core kinase domain, but only 61% identical outside it [17]. The genes encoding human AMPK-β1 and -γ1 have been localized to 12q24.1–24.3 and 12q12–q14 respectively [15].

Searching of EST and genome databases reveals at least one other isoform of β (β2) and two of γ (γ2 and γ3). The human β2 gene is at 7q35–q36 [15]. Whether these isoforms have any distinct functions is unclear at present. Both α1 and α2 readily form complexes with β1 and γ1 [9,13–15], so as yet it does not appear that specific α-subunits associate only with specific β- and γ-subunits. There are subtle differences in substrate specificity between complexes containing α1 and α2 [14]. However, the most dramatic differences between α1 and α2 observed to date are in their tissue and subcellular localization. The α1 isoform is widely distributed and appears to be almost ubiquitous. By contrast, the α2 isoform is highly expressed in liver, skeletal muscle and cardiac muscle, and at low abundance in many other cell types [17]. An even more marked difference in distribution has recently been found at the subcellular level. In INS-1 cells, a rat pancreatic β-cell line which expresses both α1 and α2, the α2 protein is almost entirely nuclear, whereas α1 is largely cytoplasmic. This was determined using isoform-specific anti-peptide antibodies, both by indirect immunofluorescence in the scanning confocal microscope and by Western blotting of cell lysates and nuclear-enriched fractions (I. Salt and D.G. Hardie, unpublished work).

The most striking findings to emerge from the cloning and sequencing of DNAs encoding the α-, β- and γ-subunits of AMPK [9–12] were the sequence similarities with the subunits of the budding-yeast (*Saccharomyces cerevisiae*) SNF1 system. AMPK-α2 is approx. 60% identical in sequence with the *SNF1* gene product within the kinase domain, and the similarity, although lower, extends through the C-terminal domain as well [8,11]. This was intriguing, because the physiological function of SNF1 was already at least partly defined from genetic studies. AMPK-γ1 is also 35% identical to the *S. cerevisiae SNF4* gene product, and there was already good evidence that the encoded protein, Snf4p, is a regulatory component of the SNF1 complex [18,19]. AMPK-β1 is related to the *S. cerevisiae* Sip1p, Sip2p and Gal83p subfamily, especially within the conserved, C-terminal ASC ('association with the SNF1 complex') domains [20] and central KIS regions [21] (see below). The implications of these homologies are discussed further below.

Structure of the yeast SNF1 complex

The *SNF1* (*sucrose non-fermenting 1*) gene (also called *CAT1* and *CCR1*) was originally defined via mutants that would grow on glucose but not on glycerol or maltose [22], ethanol [23], or sucrose or raffinose [24]. A characteristic of growth on most carbon sources other than glucose is that it requires de-repression of genes that are repressed in the presence of glucose, such as *SUC2* (encoding invertase, required to metabolize sucrose), *GAL1*, *GAL7*, *GAL10* (encoding enzymes required for galactose metabolism) and *COX6* (encoding cytochrome oxidase, required for oxidative growth on non-fermentable carbon sources). The *SNF1* gene was found to be essential for de-repression of all of these genes when glucose was removed from the medium. The gene was cloned by complementation of mutants, and found to encode a protein kinase [25], shown later to be closely related to the AMPK α-subunits [8,11].

The *SNF4* gene (also known as *CAT3*) was isolated in the same genetic screens as *SNF1/CAT1*, and is also essential for de-repression of glucose-repressed genes, although mutations in *SNF4* can be partially alleviated by overexpression of *SNF1*, or deletion of the region of *SNF1* encoding the C-terminal domain [19]. Cloning and sequencing of *SNF4* showed that it encodes a 36 kDa protein [18], later found to be related to the AMPK γ-subunit [9,10]. The products of the *SNF1* and *SNF4* genes, i.e. Snf1p and Snf4p, associate into a complex, as revealed by co-immunoprecipitation [18], two-hybrid analysis [26] and biochemical purification [8,27].

Two further SNF1-interacting proteins (Sip1p and Sip2p) were cloned from yeast by two-hybrid analysis using Snf1p as bait [28], while a third, Gal83p [20], had already been defined by other genetic screens [29]. Sip1p, Sip2p and Gal83p all co-immunoprecipitate from yeast extracts using antibodies against Snf1p, and all are phosphorylated *in vivo* in a *SNF1*-dependent manner [20]. Although none of Sip1p, Sip2p or Gal83p could be detected in stoichiometric amounts in the biochemically purified SNF1 complex [27], it seems likely that the yeast SNF1 complexes, like mammalian AMPK, are heterotrimers formed from Snf1p, Snf4p and one of Sip1p, Sip2p or Gal83p.

The interactions between Snf1p, Snf4p and Sip1p/Sip2p/Gal83p have recently been mapped in detail by two-hybrid analysis [21,30]. Snf4p interacts with the conserved C-terminal domains of Sip1p/Sip2p/Gal83p, previously termed the ASC domains. The C-terminal 'regulatory' domain of Snf1p interacts with Sip1p/Sip2p/Gal83p via a smaller segment N-terminal to the ASC domain, known as the KIS ('kinase interaction sequence') region. If the cells are grown under repressing conditions (i.e. high glucose), Snf1p and Snf4p do not interact directly, but only indirectly via the third subunit. Under these conditions the C-terminal 'regulatory' domain of Snf1p interacts with the kinase domain and appears to inhibit it. When glucose is removed from the medium, Snf4p now interacts with the regulatory domain of Snf1p, displacing the kinase domain and thus activating the kinase.

It seems very likely that this model will also be broadly applicable to AMPK. The regions that are most highly conserved between the sequences of Sip1p, Sip2p and Gal83p and the mammalian β1-subunit are the ASC and KIS segments. Woods et al. [9] reported that, after expression in reticulocyte lysates, the mammalian α-and γ-subunits did not form a stable interaction, whereas $\alpha\beta$ or $\beta\gamma$ did form stable complexes. On the other hand, Dyck et al. [13] observed stable $\alpha\gamma$ complexes in the absence of β, albeit at low levels, after transfection in COS7 cells. One explanation for this apparent discrepancy may be that the kinase is in a different state of activation in the two cases. The reticulocyte lysate system utilizes an ATP-regenerating system, which would mean that AMP levels were very low and the kinase would be inactive (resulting in, according to the yeast model, a lack of association between α and γ). On the other hand the kinase may be at least partially activated in COS7 cells, particularly in response to the stress of cell isolation prior to homogenization. Under these conditions the yeast model would suggest that the α- and γ-subunits should associate.

Regulation of mammalian AMPK

Different preparations of AMPK are allosterically activated 3–5-fold by AMP, with an EC_{50} in the standard assay (200 μM ATP) of 4 μM. Importantly, the effect is antagonized by ATP, so that at a more physiological ATP concentration of 4 mM the EC_{50} increases to 30 μM [31]. This suggests that the kinase would be activated by low cellular 'energy charge' [32], as originally suggested by Kim and co-workers [33]. However, it is now abundantly clear that activation by AMP involves more than simple allosteric regulation. AMPK is also regulated by phosphorylation by an upstream kinase, AMPK kinase (AMPKK), due in part to phosphorylation at Thr-172 [34], a site within the so-called 'activation segment' [35] where several other protein kinases are regulated (Fig. 1). As well as causing allosteric activation, 5′-AMP is also absolutely required for phosphorylation and activation by AMPKK [36], and we have shown that this is both substrate-mediated (i.e. due to AMP binding to AMPK), and enzyme-

PKA-Cα	**DFG**FAKRVKGRTW**T**LCGTPEYL**APE**
PKCα	**DFG**MCKEHMMDGVTTR**T**FCGTPDYI**APE**
Cdc2	**DFG**LARAFGIPIRVY**T**HEVVTLWYR**SPE**
MAPK	**DFG**LARIADPEHDHTGFL**T**E**Y**VATRWYR**APE**
RSK^{mo-1}	**DFG**FAKQLRAENGLLM**T**PCY**T**ANFV**APE**
MEK1	**DFG**VSGQLID**S**MAN**S**FVGTRSYM**SPE**
GSK3	**DFG**SAKQLVRGEPNVS**Y**ICSRYYR**APE**
CaMKIV	**DFG**LSKIVEHQVLMK**T**VCGTPGYC**APE**
CAMKI	**DFG**LSKMEDPGSVLS**T**ACGTPGYV**APE**
AMPKα	**DFG**LSNMMSDGEFLR**T**SCGSPNYA**APE**
Snf1p	**DFG**LSNIMTDGNFLK**T**SCGSPNYA**APE**

Fig. 1. Alignment of the activation segments of a number of protein kinases that are activated by phosphorylation. The Asp-Phe-Gly (DFG) and Ala-Pro-Glu/Ser-Pro-Glu (APE/SPE) motifs are in bold type, and the sequences are aligned on the latter. Phosphorylation sites are in bold type and underlined. PKA-Cα, human cAMP-dependent protein kinase catalytic subunit α isoform; PKCα, rat protein kinase C α isoform; Cdc2, human Cdc2; MAPK, human ERK1 (MAP kinase); RSK^{mo-1}, mouse ribosomal protein S6 kinase; MEK1, mouse MAPK/ERK kinase 1 (MAP kinase kinase); GSK3, glycogen synthase kinase 3; CaMKIV and CaMKI, mouse calmodulin-dependent protein kinases IV and I respectively; AMPKα, rat AMPK α-subunit; Snf1p, *S. cerevisiae* Snf1p. Reproduced from [3], with permission.

mediated (i.e. due to AMP binding to AMPKK) [37]. In addition, AMP binding to AMPK almost completely prevents dephosphorylation and inactivation by protein phosphatase-2C [38]. A model to account for these observations is presented in Fig. 2. Borrowing from the classical Monod/Changeux/Jacob model for allosteric enzymes [39], AMPK is envisaged as existing in more-active R conformations and less-active T conformations. Each of these also exists in phosphorylated and dephosphorylated forms, making four forms in all. AMP is proposed to stabilize the R states, thus shifting the equilibrium in favour of the R states and causing up to 5-fold activation of the phosphorylated form. More importantly, the T→R transition of the dephosphorylated form makes it a much better substrate for the upstream kinase. Phosphorylation causes at least 50-fold activation; it may be that the dephosphorylated forms are completely inactive, although this is still not certain. Dissociation of AMP from the phosphorylated R state would cause the transition to the T state, leading to loss of the 5-fold allosteric activation and, more importantly, allowing dephosphorylation and inactivation by protein phosphatase-2C.

One useful agent that mimics the effects of AMP on the system is 5-aminoimidazole-4-carboxamide ribonucleoside 5′-monophosphate, which not only causes allosteric activation of AMPK [31,40] but also promotes its phosphorylation by AMPKK [31]. If the compound is given to intact cells as the non-phosphorylated riboside (AICAriboside), it is taken up and accumulates inside the cells in the active phosphorylated form. This

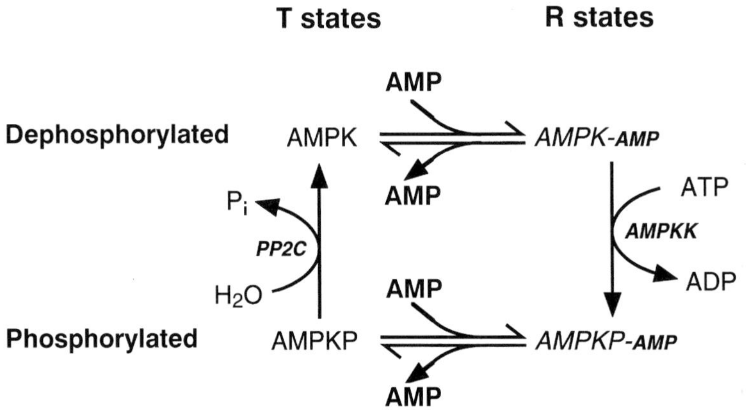

Fig. 2. Model for the regulation of AMPK by AMP. In this model, AMPK exists in four states, i.e. phosphorylated (bottom) and dephosphorylated (top) forms, and R (right) and T (left) conformations. AMP promotes formation of the R states, either by inducing the conformational change from T to R or by preferentially binding to, and stabilizing, the R states. Only the R state is phosphorylated significantly by AMPKK, whereas only the T state is significantly dephosphorylated by protein phosphatase-2C (PP2C). Reproduced from [3], with permission.

represents a method for activating AMPK in intact cells without disturbing AMP or ATP levels [31,41].

Why should AMP activate the AMPK system through multiple mechanisms? We have performed computer modelling based on the equations developed by Goldbeter and Koshland [42] to describe the behaviour of covalent modification systems. Having three effects of AMP on the phosphorylation status (activation of AMPKK, and binding to AMPK making it a better substrate for AMPKK and a worse substrate for protein phosphatase-2C) means that, for a given increase in AMP concentration over a certain critical range, the response of the system is much larger than if AMP had a single input. Another key factor that makes the system respond in a highly sensitive manner is the fact that the K_m of AMPKK for AMPK (5 nM) is extremely low, at least 30-fold lower than the estimated total concentration of AMPK in rat hepatocytes. This means that AMPKK remains saturated with its substrate and the phosphorylation continues at a maximal rate until the proportion of AMPK in the dephosphorylated form has dropped to a very low level. This phenomenon has been termed 'zero-order ultrasensitivity' by Goldbeter and Koshland [42].

All of the effects of AMP on the AMPK system appear to be antagonized by high ATP concentrations, although it is not yet clear whether the effect of ATP on phosphorylation is substrate-mediated (due to binding to AMPK) or enzyme-mediated (due to binding to AMPKK), or both [34,38]. Thus the signals that activate the system are high AMP concentration combined with low ATP concentration. In the intact cell the AMP/ATP ratio is determined mainly by the activity of adenylate kinase, which catalyses the reaction $2ADP \leftrightarrow ATP + AMP$. Adenylate kinase appears to maintain this reaction close to equilibrium at all times in eukaryotic cells, and the equilibrium constant is approximately equal to 1. It is easy to show from this that the AMP/ATP ratio would vary as the square of the ADP/ATP ratio. In a normal, fully energized cell, the ATP/ADP ratio is maintained by oxidative phosphorylation at something of the order of 10:1. Under these conditions the ATP/AMP ratio would be of the order of 100:1 and the AMPK system would be essentially inactive. If, however, some adverse treatment of the cell was to interfere with ATP production, a 3-fold increase in the ADP/ATP ratio would result in a 9-fold increase in the AMP/ATP ratio. The latter ratio therefore represents a sensitive indicator that the energy status of the cell is compromised, and an increase in this ratio is the trigger that switches on the AMPK system.

We have therefore developed the concept that the AMPK system is like a cellular 'fuel gauge' [3] which is activated by a low energy status and which initiates protective measures to either conserve ATP or promote alternative methods of ATP regeneration. An even better analogy is with the hardware and software components in laptop computers which monitor the state of the battery charge and initiate emergency measures when it is low. These measures include giving warning messages requesting that the user provides an alternative energy source, and also activation of energy-conserving events such as dimming of the screen.

What conditions do cause the AMPK system to be activated? A number of different environmental stresses compromise the energy status of eukaryotic cells, and thus cause a large activation of AMPK. These include heat shock and the poison arsenite (which inhibits the tricarboxylic acid cycle) in isolated hepatocytes [43], ischaemia in cardiac muscle [44], and exercise in skeletal muscle [45,46]. Activation of AMPK both conserves ATP by inhibiting biosynthetic pathways and also promotes at least one alternative pathway of ATP production (fatty acid oxidation). The detailed mechanisms of these downstream events will be discussed in a later section.

Regulation of SNF1

Until recently the regulation of SNF1 was poorly understood at the molecular level, although the genetic studies had suggested that its function was required to de-repress glucose-repressed genes when glucose was removed from the medium. The finding, discussed above, that the SNF1 complex is closely related to AMPK suggested that it might be possible to assay the activity of SNF1 using the SAMS peptide, a synthetic peptide based on part of the sequence of rat acetyl-CoA carboxylase which was developed for the assay of AMPK [47]. This was indeed feasible, and it was shown that SNF1 is essentially inactive when glucose is present in the medium, but that it was dramatically activated on removal of glucose [27,48]. The activation was due to phosphorylation, since it could be reversed by treatment with a homogeneous preparation of protein phosphatase-2A [27]. The whole kinase cascade appears to be highly conserved between yeast and mammals, since mammalian AMPKK could be used to re-activate the inactive, dephosphorylated form of yeast SNF1 [27,48]. In addition, mutation of Thr-210 in Snf1p (equivalent to the major site, Thr-172, at which AMPKK phosphorylates AMPK) to alanine completely abrogates the function of SNF1 *in vivo* [49].

Although the phenomenon of glucose repression was first described in yeast as long ago as 1900 [50], the identity of the key intracellular signals that initiate de-repression of glucose-repressed genes has remained enigmatic. The many similarities between the mammalian and yeast systems raise the intriguing possibility that increases in AMP and decreases in ATP might be the long-sought signals. Consistent with this idea, removal of glucose from yeast growing exponentially in high glucose concentrations causes massive increases in AMP and decreases in ATP (the AMP/ATP ratio increased >200-fold within 5 min). Under a variety of different conditions, there seems to be a good correlation between AMP and ATP levels *in vivo* and the activation state of SNF1 measured in an extract prepared from the cells [27]. The problem with this hypothesis is that, at the time of writing, no effect of AMP on activation of the SNF1 system has been observed in cell-free assays. Certainly AMP does not seem to cause allosteric activation of SNF1 [27,48]. However, the possibility that AMP promotes phosphorylation and activation of SNF1 by the upstream protein kinase is currently difficult to test, because the cell-free systems available are very crude and are contaminated with activities that generate AMP during the assay (probably adenylate kinase) [27].

As discussed in a previous section, recent two-hybrid analysis of the interactions between Snf1p and Snf4p [30] suggests that the kinase domain of Snf1p is maintained in an inactive state by an inhibitory association with the C-terminal domain of the same subunit. Through mechanisms which remain unclear, but could perhaps involve binding of AMP, this interaction is somehow disrupted when glucose is removed from the medium, because Snf4p now interacts with the inhibitory region of Snf1p instead [30]. An obvious hypothesis would be that the Thr-210 site in the kinase domain is now exposed and available to be phosphorylated by the upstream kinase. The active, phosphorylated kinase domain could then phosphorylate downstream targets. It has been suggested that one function of the Sip1p/Sip2p/Gal83p subunits (equivalent to the mammalian AMPK β-subunits) might be to target the active complex to specific cellular locations, perhaps via their non-conserved N-terminal domains [20]. The principal evidence for this is that specific point mutations in *GAL83* lead to partially constitutive expression of the *GAL* genes, but not of other genes that are targets for the SNF1 complex [29].

Genetic studies have provided certain other clues to the mechanisms of activation of SNF1. Mutations in several genes yield constitutive expression of glucose-repressed genes. These genes appear to act upstream of *SNF1*, and the implication is that loss of their function causes SNF1 to be active even though glucose is present in the medium. One of these is *HXK2* (*HEX1*), which encodes the major PII isoform of hexokinase. This observation could be reconciled with the hypotheses discussed above if one was to assume that a substantial defect in hexokinase activity led to a lower rate of glycolysis, and hence decreased cellular ATP and increased AMP. Two other genes with the same phenotype are *GLC7* and *REG1*, which encode an isoform of protein phosphatase-1 [51] and a putative targeting/regulatory subunit of the *GLC7* gene product [52] respectively. These phenotypes are consistent with the hypothesis that the Glc7p/Reg1p complex is the phosphatase normally responsible for dephosphorylation of SNF1, although other interpretations are possible.

Targets of AMPK

AMPK was discovered independently in two laboratories in 1973 due to its ability to inactivate 3-hydroxy-3-methylglutaryl-CoA (HMG-CoA) reductase [5] and acetyl-CoA carboxylase [6], although not until 1987 was it realized that these were functions of a single protein kinase [53]. HMG-CoA reductase and acetyl-CoA carboxylase remain the best-established physiological substrates. The sites phosphorylated by AMPK *in vitro* have been characterized (Ser-871 in HMG-CoA reductase; Ser-79, Ser-1200 and Ser-1215 in acetyl-CoA carboxylase [54–56]) and all of these sites are phosphorylated *in vivo* or in intact cells under conditions where the kinase is activated [57–59]. The two pathways regulated by the kinase, i.e. sterol synthesis (HMG-CoA reductase) and fatty acid synthesis (acetyl-CoA carboxylase), are also dramatically inactivated by treatments that activate the kinase, such as heat shock or arsenite [43]. Sato et al. [60] have also shown that the inactivation of sterol synthesis induced by treatment of cells expressing wild-type HMG-CoA reductase with 2-deoxyglucose (which acts as a phosphate trap and thus depletes ATP) is completely abolished in cells expressing a Ser-871 → Ala mutant.

The inhibition of these two biosynthetic pathways (sterol and fatty acid synthesis) in cells depleted of ATP would act as a protective mechanism by conserving ATP (and NADPH). Phosphorylation of acetyl-CoA carboxylase has an additional effect, i.e. production of more ATP by activation of fatty acid oxidation. The product of acetyl-CoA carboxylase, malonyl-CoA, is a potent inhibitor of carnitine palmitoyltransferase I, the enzyme involved in the initial uptake of fatty acids into mitochondria, which is thought to exert significant control over the whole pathway. Activation of AMPK therefore causes inactivation of acetyl-CoA carboxylase, lowering of the malonyl-CoA concentration and activation of fatty acid oxidation. This mechanism has been demonstrated to operate in isolated hepatocytes [61], skeletal muscle [45,62] and cardiac muscle [44,63].

A third physiological target for AMPK is hormone-sensitive lipase, the enzyme responsible for the breakdown of triacylglycerol in adipocytes and of cholesterol esters in many other cell types [64]. Hormone-sensitive lipase is activated by cAMP-dependent protein kinase by phosphorylation at Ser-563, and this appears to be the mechanism by which triacylglycerol and cholesterol esters are broken down in response to cAMP-elevating hormones [65]. AMPK phosphorylates a neighbouring serine residue (Ser-565) and, although this has no effect on its own, it prevents phosphorylation and activation at Ser-563 [66]. Ser-565 is phosphorylated in intact cells [67], and activation of AMPK in isolated adipocytes antagonizes the effect of cAMP-elevating agents on triacylglycerol breakdown [31,68]. We have argued [3] that this is a mechanism for conserving ATP, in that it may prevent recycling of fatty acids into triacylglycerols under conditions where they are not removed sufficiently rapidly from the cells. Recycling of fatty acids is a 'futile' cycle which consumes ATP.

Although all of these well-established physiological substrates are involved in lipid metabolism, we believe that there may be many other targets that are involved in other processes. AMPK phosphorylates and inactivates the muscle isoform of glycogen synthase *in vitro* [69]. Although it is not yet clear that this occurs *in vivo*, this would make physiological sense, since glycogen synthesis consumes ATP and UTP. Another target in cell-free assays is the protein kinase Raf-1, which is phosphorylated at Ser-621 (and at another site, probably Ser-259). Raf-1 is a key component in the signal transduction cascade which couples the binding of growth factors to receptors of the protein-tyrosine kinase class, and activation of the classical mitogen-activated protein kinase (MAP kinase) cascade [70]. Kolch's group [71] have shown that overexpression of the catalytic subunit of cAMP-dependent protein kinase, which also leads to phosphorylation of Ser-621, inhibited basal and epidermal-growth-factor-stimulated Raf-1 activity in COS cells [71]. Consistent with this, we have recently found that activation of AMPK in isolated rat hepatocytes using AICAriboside led to inhibition of the basal and epidermal-growth-factor-stimulated activities of both Raf-1 and its downstream partner, MAP kinase (J.A. Corton and D.G. Hardie, unpublished work). If confirmed, this would represent an intriguing mechanism whereby cellular stresses that remove ATP would inhibit the transition of cells from the quiescent to the dividing state.

In view of the abundant evidence that the yeast homologue SNF1 is involved in regulation of gene expression, it seems likely that this will also be true for mammalian AMPK. Another interesting target identified recently in cell-free assays is the transcription factor CREB (cAMP response element binding protein), which is phosphorylated at Ser-133 and one other site (A. Clifton and D.G. Hardie, unpublished work). Although it remains to be proven that this occurs in intact cells, the finding that the α2 isoform of AMPK is largely nuclear (see above) makes transcription factors particularly interesting candidates as targets.

Downstream targets for the SNF1 complex

Few substrates of SNF1 have been identified by direct biochemical analysis as yet, although yeast acetyl-CoA carboxylase is phosphorylated and inactivated by SNF1 both *in vivo* [48] and *in vitro* [8]. Acetyl-CoA carboxylase is therefore one physiological target that is shared by the yeast and mammalian homologues. Most other putative targets have currently been identified only genetically, i.e. as gene products that appear to act downstream of *SNF1*. These include a number of putative transcription factors, such as Mig1p [72], Sip4p [73] and Msn2p [74]. Mig1p is a particularly interesting candidate as a direct target for the SNF1 complex. In the presence of glucose, Mig1p is involved in the repression of transcription of several genes, probably because it recruits the co-repressors Ssn6p and Tup1p (which are not themselves DNA-binding proteins) to the promoters [72]. Deletion analysis has shown that a regulatory region of Mig1p outside the DNA-binding region, known as the R1 domain, is required for SNF1-mediated relief from repression [75]. We have recently expressed Mig1p in bacteria, and shown that it is an excellent substrate for the purified SNF1 complex in cell-free assays (D. Carling and D.G. Hardie, unpublished work). Intriguingly, mapping reveals that there are four phosphorylation sites, all of which lie within or very close to the R1 domain. The significance of these sites to *MIG1* function is now being addressed by expressing point mutants *in vivo*.

Conclusions and perspectives

The finding of the striking homology between the mammalian AMPK and yeast SNF1 complexes has had a major impact on the study of both systems. Since the studies on AMPK had been predominantly biochemical in nature, whereas those on SNF1 had utilized genetics, the type of information available for the two systems was markedly different. AMPK had been implicated in the response to environmental stress, and the mechanisms by which it is activated in response to stress were well established. Its downstream targets appear to be predominantly metabolic enzymes, particularly those involved in lipid metabolism. The yeast SNF1 system, on the other hand, is involved in the response to glucose starvation (i.e. nutritional stress). The mechanisms by which it is activated were poorly understood (in fact, it was not even clear that it **was** activated [25]). A number of downstream-acting genes had been identified, but it was unclear whether any of these are direct substrates for the kinase. Most of the downstream-acting genes appear to encode transcription factors rather than metabolic enzymes, and the genes that are regulated by these factors are involved in carbohydrate rather than lipid metabolism.

Although it remains to be seen to what extent mammalian AMPK can be used as a valid model for yeast SNF1, and vice versa, we believe that these will be powerful approaches. Some new insights have already emerged from applying these strategies, e.g. (1) the finding that SNF1 is activated by phosphorylation in response to glucose starvation, and that this correlates with large changes in adenine nucleotides [27]; and (2) the finding that one form of AMPK is nuclear and that it phosphorylates a transcription factor (CREB), at least *in vitro*. Although the regulation and physiological functions of AMPK and SNF1 may currently appear to be rather different, we believe that the underlying mechanisms will turn out to be highly conserved.

Studies in the author's laboratory have been supported by the Wellcome Trust and the Medical Research Council. This review describes work performed by several members of the laboratory, including Julia Corton, Steve Davies, Simon Hawley, Ian Salt and Wayne Wilson. I also thank David Norman, Alan Prescott and Angus Lamond (University of Dundee), David Carling (Royal Postgraduate Medical School, London) and Will Winder (Brigham Young University, Provo, Utah, U.S.A.) for enjoyable collaborative studies.

References

1. Hardie, D.G. (1994) Cell. Signalling **6**, 813–821
2. Hardie, D.G., Carling, D. and Halford, N.G. (1994) Semin. Cell Biol. **5**, 409–416
3. Hardie, D.G. and Carling, D. (1997) Eur. J. Biochem. **246**, 259–273
4. Hardie, D.G. (1992) Biochim. Biophys. Acta **1123**, 231–238

5. Beg, Z.H., Allmann, D.W. and Gibson, D.M. (1973) Biochem. Biophys. Res. Commun. **54**, 1362–1369
6. Carlson, C.A. and Kim, K.H. (1973) J. Biol. Chem. **248**, 378–380
7. Davies, S.P., Hawley, S.A., Woods, A., Carling, D., Haystead, T.A.J. and Hardie, D.G. (1994) Eur. J. Biochem. **223**, 351–357
8. Mitchelhill, K.I., Stapleton, D., Gao, G., House, C., Michell, B., Katsis, F., Witters, L.A. and Kemp, B.E. (1994) J. Biol. Chem. **269**, 2361–2364
9. Woods, A., Cheung, P.C.F., Smith, F.C., Davison, M.D., Scott, J., Beri, R.K. and Carling, D. (1996) J. Biol. Chem. **271**, 10282–10290
10. Gao, G., Fernandez, S., Stapleton, D., Auster, A.S., Widmer, J., Dyck, J.R.B., Kemp, B.E. and Witters, L.A. (1996) J. Biol. Chem. **271**, 8675–8681
11. Carling, D., Aguan, K., Woods, A., Verhoeven, A.J.M., Beri, R.K., Brennan, C.H., Sidebottom, C., Davison, M.D. and Scott, J. (1994) J. Biol. Chem. **269**, 11442–11448
12. Gao, G., Widmer, J., Stapleton, D., Teh, T., Cox, T., Kemp, B.E. and Witters, L.A. (1995) Biochim. Biophys. Acta **1266**, 73–82
13. Dyck, J.R.B., Gao, G., Widmer, J., Stapleton, D., Fernandez, C.S., Kemp, B.E. and Witters, L.A. (1996) J. Biol. Chem. **271**, 17798–17803
14. Woods, A., Salt, I., Scott, J., Hardie, D.G. and Carling, D. (1996) FEBS Lett. **397**, 347–351
15. Stapleton, D., Woollatt, E., Mitchelhill, K.I., Nicholl, J.K., Fernandez, C.S., Michell, B.J., Witters, L.A., Power, D.A., Sutherland, G.R. and Kemp, B.E. (1997) FEBS Lett. **409**, 452–456
16. Beri, R.K., Marley, A.E., See, C.G., Sopwith, W.F., Aguan, K., Carling, D., Scott, J. and Carey, F. (1994) FEBS Lett. **356**, 117–121
17. Stapleton, D., Mitchelhill, K.I., Gao, G., Widmer, J., Michell, B.J., Teh, T., House, C.M., Fernandez, C.S., Cox, T., Witters, L.A. and Kemp, B.E. (1996) J. Biol. Chem. **271**, 611–614
18. Celenza, J.L., Eng, F.J. and Carlson, M. (1989) Mol. Cell. Biol. **9**, 5045–5054
19. Celenza, J.L. and Carlson, M. (1989) Mol. Cell. Biol. **9**, 5034–5044
20. Yang, X., Jiang, R. and Carlson, M. (1994) EMBO J. **13**, 5878–5886
21. Jiang, R. and Carlson, M. (1997) Mol. Cell. Biol. **17**, 2099–2106
22. Zimmermann, F.K., Kaufmann, I., Rasenberger, H. and Haussman, P. (1977) Mol. Gen. Genet. **151**, 95–103
23. Ciriacy, M. (1977) Mol. Gen. Genet. **154**, 213–220
24. Neigeborn, L. and Carlson, M. (1984) Genetics **108**, 845–858
25. Celenza, J.L. and Carlson, M. (1986) Science **233**, 1175–1180
26. Fields, S. and Song, O.K. (1989) Nature (London) **340**, 245–246
27. Wilson, W.A., Hawley, S.A. and Hardie, D.G. (1996) Curr. Biol. **6**, 1426–1434
28. Yang, X., Hubbard, E.J. and Carlson, M. (1992) Science **257**, 680–682
29. Erickson, J.R. and Johnston, M. (1993) Genetics **135**, 655–664
30. Jiang, R. and Carlson, M. (1996) Genes Dev. **10**, 3105–3115
31. Corton, J.M., Gillespie, J.G., Hawley, S.A. and Hardie, D.G. (1995) Eur. J. Biochem. **229**, 558–565
32. Atkinson, D.E. (1977) Cellular Energy Metabolism and its Regulation, Academic Press, New York
33. Yeh, L.A., Lee, K.H. and Kim, K.H. (1980) J. Biol. Chem. **255**, 2308–2314
34. Hawley, S.A., Davison, M., Woods, A., Davies, S.P., Beri, R.K., Carling, D. and Hardie, D.G. (1996) J. Biol. Chem. **271**, 27879–27887
35. Johnson, L.N., Noble, M.E.M. and Owen, D.J. (1996) Cell **85**, 149–158
36. Moore, F., Weekes, J. and Hardie, D.G. (1991) Eur. J. Biochem. **199**, 691–697

37. Hawley, S.A., Selbert, M.A., Goldstein, E.G., Edelman, A.M., Carling, D. and Hardie, D.G. (1995) J. Biol. Chem. **270**, 27186–27191
38. Davies, S.P., Helps, N.R., Cohen, P.T.W. and Hardie, D.G. (1995) FEBS Lett. **377**, 421–425
39. Monod, J., Changeux, J.P. and Jacob, F. (1963) J. Mol. Biol. **6**, 306–329
40. Henin, N., Vincent, M.F. and Van den Berghe, G. (1996) Biochim. Biophys. Acta **1290**, 197–203
41. Henin, N., Vincent, M.F., Gruber, H.E. and Van den Berghe, G. (1995) FASEB J. **9**, 541–546
42. Goldbeter, A. and Koshland, D.E. (1981) Proc. Natl. Acad. Sci. U.S.A. **78**, 6840–6844
43. Corton, J.M., Gillespie, J.G. and Hardie, D.G. (1994) Curr. Biol. **4**, 315–324
44. Kudo, N., Barr, A.J., Barr, R.L., Desai, S. and Lopaschuk, G.D. (1995) J. Biol. Chem. **270**, 17513–17520
45. Winder, W.W. and Hardie, D.G. (1996) Am. J. Physiol. **270**, E299–E304
46. Vavvas, D., Apazidis, A., Saha, A.K., Gamble, J., Patel, A., Kemp, B.E., Witters, L.A. and Ruderman, N.B. (1997) J. Biol. Chem. **272**, 13255–13261
47. Davies, S.P., Carling, D. and Hardie, D.G. (1989) Eur. J. Biochem. **186**, 123–128
48. Woods, A., Munday, M.R., Scott, J., Yang, X., Carlson, M. and Carling, D. (1994) J. Biol. Chem. **269**, 19509–19515
49. Estruch, F., Treitel, M.A., Yang, X. and Carlson, M. (1992) Genetics **132**, 639–650
50. Dienert, F. (1900) Ann. Inst. Pasteur **19**, 139–189
51. Tu, J. and Carlson, M. (1994) Mol. Cell. Biol. **14**, 6789–6796
52. Tu, J. and Carlson, M. (1995) EMBO J. **14**, 5939–5946
53. Carling, D., Zammit, V.A. and Hardie, D.G. (1987) FEBS Lett. **223**, 217–222
54. Clarke, P.R. and Hardie, D.G. (1990) EMBO J. **9**, 2439–2446
55. Munday, M.R., Campbell, D.G., Carling, D. and Hardie, D.G. (1988) Eur. J. Biochem. **175**, 331–338
56. Davies, S.P., Sim, A.T.R. and Hardie, D.G. (1990) Eur. J. Biochem. **187**, 183–190
57. Sim, A.T.R. and Hardie, D.G. (1988) FEBS Lett. **233**, 294–298
58. Davies, S.P., Carling, D., Munday, M.R. and Hardie, D.G. (1991) Eur. J. Biochem. **203**, 615–623
59. Gillespie, J.G. and Hardie, D.G. (1992) FEBS Lett. **306**, 59–62
60. Sato, R., Goldstein, J.L. and Brown, M.S. (1993) Proc. Natl. Acad. Sci. U.S.A. **90**, 9261–9265
61. Velasco, G., Geelen, M.J.H. and Guzman, M. (1997) Arch. Biochem. Biophys. **337**, 169–175
62. Hutber, C.A., Hardie, D.G. and Winder, W.W. (1997) Am. J. Physiol. **272**, E262–E266
63. Kudo, N., Gillespie, J.G., Kung, L., Witters, L.A., Schulz, R., Clanachan, A.S. and Lopaschuk, G.D. (1996) Biochim. Biophys. Acta **1301**, 67–75
64. Yeaman, S.J. (1990) Biochim. Biophys. Acta. **1052**, 128–132
65. Garton, A.J., Campbell, D.G., Cohen, P. and Yeaman, S.J. (1988) FEBS Lett. **229**, 68–72
66. Garton, A.J., Campbell, D.G., Carling, D., Hardie, D.G., Colbran, R.J. and Yeaman, S.J. (1989) Eur. J. Biochem. **179**, 249–254
67. Garton, A.J. and Yeaman, S.J. (1990) Eur. J. Biochem. **191**, 245–250
68. Sullivan, J.E., Brocklehurst, K.J., Marley, A.E., Carey, F., Carling, D. and Beri, R.K. (1994) FEBS Lett. **353**, 33–36
69. Carling, D. and Hardie, D.G. (1989) Biochim. Biophys. Acta **1012**, 81–86
70. Morrison, D.K. and Cutler, R.E. (1997) Curr. Opin. Cell Biol. **9**, 174–179

71. Mischak, H., Seitz, T., Janosch, P., Eulitz, M., Steen, H., Schellerer, M., Philipp, A. and Kolch, W. (1996) Mol. Cell. Biol. **16**, 5409–5418
72. Treitel, M.A. and Carlson, M. (1995) Proc. Natl. Acad. Sci. U.S.A. **92**, 3132–3136
73. Lesage, P., Yang, X. and Carlson, M. (1996) Mol. Cell. Biol. **16**, 1921–1928
74. Estruch, F. and Carlson, M. (1993) Mol. Cell. Biol. **13**, 3872–3881
75. Ostling, J., Carlberg, M. and Ronne, H. (1996) Mol. Cell. Biol. **16**, 753–761

Biochem. Soc. Symp. **64**, 29–48
Printed in Great Britain

3

Making the connection: coupling of stress-activated ERK/MAPK (extracellular-signal-regulated kinase/mitogen-activated protein kinase) core signalling modules to extracellular stimuli and biological responses

John M. Kyriakis

Diabetes Research Laboratory, Massachusetts General Hospital and Department of Medicine, Harvard Medical School, M.G.H. East, 149 13th Street, Charlestown, MA 02129, U.S.A.

Abstract

Signal-transduction pathways that employ members of the extracellular signal-regulated kinase (ERK)/mitogen-activated protein kinase (MAPK) family of protein Ser/Thr kinases are widely conserved among eukaryotes. The multiplicity of these pathways allows the cell to respond to divergent extracellular stimuli by initiating a broad array of responses ranging from cell growth to apoptosis. ERK/MAPK pathways are comprised of a three-tiered core-signalling module wherein ERK/MAPKs are regulated by MAPK/ERK kinases (MEKs) and MEKs, in turn, are regulated by MAPK kinase kinases (MAPKKKs). The regulation of MAPKKK→MEK→ERK/MAPK core-signalling modules by upstream components is poorly understood. Mammalian stress-activated ERK/MAPK pathways have been implicated in numerous important physiological functions, including inflammatory responses and apoptosis. In this review, I will discuss how mammalian stress-regulated ERK/MAPK core-signalling modules couple with members of the SPS1 family of protein kinases and to other upstream elements, and how these stress-regulated pathways influence cell function.

Introduction

Signal transduction networks which culminate in the activation of extracellular-signal-regulated kinases (ERKs)/mitogen-activated protein kinases (MAPKs) have been widely conserved in eukaryotic cell evolution (Fig. 1) [1,2]. At the heart of these networks are so-called 'core signalling modules', consisting of the ERKs/MAPKs that are activated by concomitant Tyr and Thr phosphorylation catalysed by members of the MAPK/ERK kinase (MEK) family. MEKs, in turn, are activated by Ser/Thr phosphorylation catalysed by protein kinases of several families collectively termed MAPK kinase kinases (MAPKKKs) [1,2]. The diversity of kinases with MAPKKK activity reflects the breadth of extracellular stimuli that can activate ERK/MAPK networks. The ERKs/MAPKs themselves have a broad substrate specificity and can thus regulate an extensive range of cellular processes, including gene transcription, cytoskeletal organization, metabolite homoeostasis, cell growth and apoptosis [1–3].

While considerable progress has been made in the identification of the molecular components and regulatory relationships that comprise ERK/MAPK core signalling modules, considerably less is known of how core signalling modules are linked to events at the cell surface or to biological responses. The purpose of this review is to discuss what is known of the regulatory mechanisms upstream of mammalian stress-

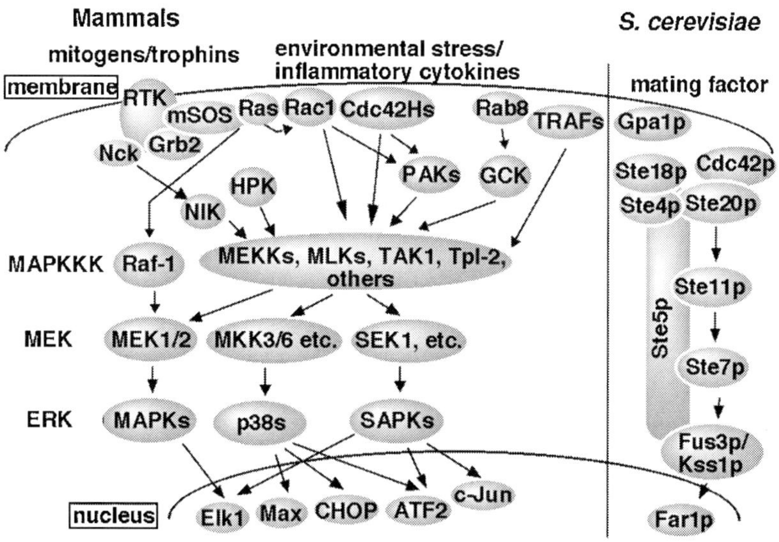

Fig. 1. ERK signalling in eukaryotes. A comparison of mammalian and yeast signalling pathways and an illustration of the concept of the core signalling module are emphasized. Only the yeast mating factor pathway is shown. For all pathways, nuclear substrates are selectively illustrated. Abbreviations: RTK, receptor tyrosine kinase; CHOP, cAMP response element-binding protein homologous protein.

regulated MAPKKKs and the biological functions of mammalian stress-activated ERK/MAPK pathways. In particular I will focus on two proximal elements implicated in the regulation of MAPKKK→MEK→ERK core modules: (1) an emerging class of protein kinases, the SPS1 family, and (2) adaptor polypeptides which associate with receptors of the tumour necrosis factor (TNF) receptor family. I will also discuss recent evidence concerning the role of mammalian stress-activated ERK/MAPK pathways in the regulation of cell cycle progression and apoptosis.

The SAPK and p38 pathways: mammalian ERK/MAPK signalling networks activated by stress and inflammation

The stress-activated protein kinases [SAPKs; also referred to as Jun N-terminal kinases (JNKs)] and the p38 MAPKs are mammalian ERK/MAPK subfamilies which are potently and preferentially activated by a variety of environmental stresses (ionizing radiation, heat shock, oxidative stress, osmotic shock), inflammatory cytokines and receptor systems of the TNF family [TNF, interleukin-1 (IL-1), CD40], and vasoactive stresses associated with the response to ischaemia (reperfusion injury, angiotensin II, endothelin). In most instances, the SAPKs and p38s are poorly activated by mitogenic stimuli coupled to Ras (Fig. 2) [1,3–9].

The SAPKs are encoded by at least three genes which are further diversified by alternative mRNA splicing into up to 12 isoforms [4,10]. The SAPKs were originally identified as the major Ser/Thr kinases respon-

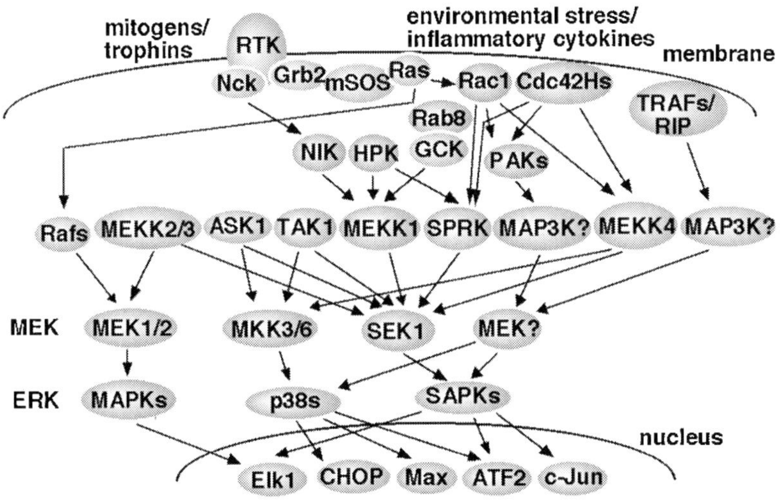

Fig. 2. ERK signalling from the membrane to the nucleus. Mammalian stress and mitogen-activated ERK/MAPK core signalling modules and their connections with upstream signalling components and nuclear targets are shown. The complex relationships among MAPKKKs (MAP3Ks) and MEKs are noteworthy. Abbreviations: RTK, receptor tyrosine kinase; CHOP, CCAAT-enhancer-binding protein homologous protein.

sible for the phosphorylation of the c-Jun transcription factor [4,5]. c-Jun dimerizes with members of the Fos or activating transcription factor (ATF) family of transcription factors to form the activator protein-1 (AP-1) transcription factor complex. AP-1 activity is induced by a number of stressful stimuli, especially TNF and UV radiation [11–13]. Much of this activation can be attributed to phosphorylation of Ser-63 and Ser-73 in the c-Jun *trans*-activation domain, which is catalysed by the SAPKs [1,3–5,14]. This phosphorylation correlates well with ligand-induced activation of c-Jun *trans*-activating activity [3,11].

Four p38 genes have been identified. The p38s are mammalian homologues of *Saccharomyces cerevisiae HOG1* [2,6–8,15,16]. The p38 family (α and β isoforms) have been implicated in the regulation of Hsp25/Hsp27 phosphorylation (where Hsp denotes heat-shock protein) via p38-catalysed phosphorylation and activation of MAPK-activated protein kinase-2 (MAPKAPK-2), the major Hsp25/Hsp27 kinase [7,8]. All p38 isoforms, especially p38-γ and -δ, can potently phosphorylate ATF2 [15,16]. However, inasmuch as the SAPKs can also phosphorylate ATF2 [3], it is unclear which family of kinases represents the physiological ATF2 kinases. Recently, p38-α was shown to phosphorylate and activate the *trans*-activating activity of the transcription factor myocyte enhancer factor-2C. Once activated, this factor, in turn, can induce expression of the c-*jun* gene in response to extracellular stimuli. By this process, p38 can also regulate AP-1 activation [17].

Several MEKs have been identified as upstream activators of the SAPKs and p38s (Fig. 2). These MEKs display a remarkable cell-type- and stimulus-dependent regulation. The SAPKs can be activated *in vitro* and *in situ* by SAPK/ERK kinase 1 (SEK1) [also known as MAPK kinase 4 (MKK4)] [18,19]. Gene disruption studies, as well as biochemical studies, indicate that SEK1 is strongly recruited by osmotic shock, reperfusion injury and the protein synthesis inhibitor anisomycin. In some systems, SEK1 is also activated by TNF [18–21]. Two other MEK activities, SAPK kinase 4 (SAPKK4) and SAPKK5, have been characterized biochemically as SAPK-specific activators. SAPKK4 and SAPKK5 are chromatographically distinct from SEK1 and appear to be particularly responsive to interleukin-1 (IL-1) stimulation [22].

MKK3 and MKK6 are p38-specific MEKs (Fig. 2) [19,23]. MKK6 appears to be the dominant p38-activating MEK identified thus far. It is activated by all known stimuli which recruit p38, and possesses significantly greater potency as a p38 activator than does MKK3 [23–25]. SEK1 can also weakly activate p38 but, under initial rate conditions, it does so at only 10% of the rate at which it activates SAPK [19,23].

A large and diverse array of protein Ser/Thr kinases has been implicated as MAPKKKs upstream of the SAPKs and p38s (Fig. 2). Such complexity of regulation is consistent with the heterogeneous nature of the extracellular stimuli that recruit the SAPKs and p38s. These include the MEK kinases [MEKK1–MEKK4 and apoptosis-stimulating kinase-1 (ASK1)], which are mammalian homologues of *S. cerevisiae STE11*

[26–32], the mixed-lineage kinases (MLKs), which share sequence identity with both Ser/Thr and Tyr kinases [33,34], transforming growth factor-β (TGF-β)-activated kinase-1 (TAK1) [35] and Tpl-2, the product of the *cot* proto-oncogene [36].

The MAPKKKs upstream of the SAPKs and p38s differ widely in the spectrum of MEKs which they can activate *in situ* and *in vitro*. Thus MEKK1, MLK2 and Src homology-3 (SH3)-domain-containing, proline-rich kinase (SPRK)/MLK3 can activate SEK1 and appear to be largely SAPK-pathway-specific [27,33,34,37]. MEKK2 and MEKK3, as well as Tpl-2, have a broader specificity. Each can activate SEK1 and the SAPK pathway *in vitro* and *in situ* [28,29,36]. In addition, each can activate the MAPK pathway *in situ*. MEKK2 and Tpl-2 do this via activation of MEK1. The mechanism of activation of the MAPK pathway by MEKK3 is unknown [28,29,36]. None of Tpl-2, MEKK2 or MEKK3 can activate co-expressed p38; however, MEKK3 can activate MKK3 *in vitro* and *in situ* [28,29,36,38]. The significance of this inconsistency remains to be determined. ASK1, MEKK4 [also called MAP three kinase-1 (MTK1)] and TAK1 can activate SEK1, MKK3 and MKK6 *in vitro*, and can activate both the SAPK and p38 pathways *in situ* [25,30–32,35]. TAK1 is thought to couple TGF-β to the SAPKs and p38s, while ASK1 appears to be activated by TNF [25,32]. MEKK4/MTK1 probably acts downstream of certain environmental stresses, insofar as high salt, anisomycin and UV activation of the p38 pathway, but not the SAPK pathway, can be blocked effectively by overexpression of kinase-inactive MEKK4/MTK1 [31].

Small G-proteins and mammalian stress signalling pathways

The critical importance of Ras superfamily GTPases in mammalian ERK/MAPK signal transduction was first appreciated with the discovery that oncogenic Ras could activate the mitogenic MAPK pathway [39,40]. All Ras family proteins act as molecular switches, relaying signals generated at the receptor to numerous intracellular effectors. Ras family proteins are active in the GTP-bound state and inactive in the GDP-bound state. All proteins of the Ras superfamily possess a slow, intrinsic, inactivating GTPase activity, which is greatly enhanced by GTPase-activating proteins, and a slow intrinsic ability to exchange GDP for GTP which is greatly accelerated by guanine nucleotide exchange factors (GEFs) [39,40]. Activation of Ras by mitogens involves the recruitment of a complex of the GEF mammalian son of sevenless (mSOS) and the SH2/SH3 adaptor Grb2 (Figs. 1 and 2). The Grb2 SH3 domains constitutively bind polyproline SH3 binding motifs on mSOS. Upon mitogen stimulation, the Grb2 SH2 domain binds phosphotyrosine residues present on autophosphorylated receptor and non-receptor tyrosine kinases. By this process Grb2–mSOS is brought to the membrane. Once at the membrane, where Ras resides, mSOS promotes GDP/GTP exchange on Ras and thus Ras activation. GTP–Ras then binds its effectors, among them MAPKKKs of the Raf family [40].

Ras superfamily GTPases of the Rho family (the Rhos, Racs and Cdc42s) were originally thought only to regulate the actin cytoskeleton [41,42]. Recently, however, these GTPases have also been implicated in ERK/MAPK signal transduction (Figs. 1 and 2); however, their regulation and effectors are less well defined. Constitutively active mutants of the Rho family GTPases Rac1 and Cdc42Hs were shown to activate co-expressed SAPK and p38 [43–45]. Rac1 itself may be regulated by Ras, inasmuch as oncogenic Ras can induce changes in the cortical cytoskeleton, such as membrane ruffling, which can also be induced by constitutively active Rac1. Moreover, dominant inhibitory Rac1 mutants can block Ras-induced cytoskeletal effects and inhibit Ras transformation, implicating Rac1 in Ras-regulated growth [46,47]. On the other hand, the function of Cdc42Hs is less well understood. In Swiss 3T3 cells, dominant inhibitory Cdc42Hs can inhibit cell-cycle entry from G0 [48]. Conversely, as described below, we have observed that Cdc42Hs can inhibit serum-stimulated NIH3T3 cell G1/S progression once cells are in G1 [49].

Downstream targets of Rac1 and Cdc42Hs all appear to possess a common binding motif which is critical for G-protein binding [50]. This domain, the Cdc42/Rac interaction and binding domain (CRIB domain), is present on several protein kinases which have been shown to lie upstream of the SAPKs and p38s [30,50,51]. The first such kinase to be identified was p21-activated kinase-1 (PAK1). PAK1 is a member of an emerging family of Ser/Thr kinases closely related to yeast *STE20* (Fig. 1) [2,52]. These kinases contain an N-terminal regulatory domain, which includes the CRIB motif, and a C-terminal kinase domain. All mammalian PAKs bind Rac1 and Cdc42Hs, and this binding is necessary for activation of PAK kinase activity [51,52].

S. cerevisiae Ste20p has been shown to play a role in the regulation of the pheromone-response ERK/MAPK pathway (Fig. 1) [2]; it was for this reason that the PAKs were thought to be important regulators of mammalian ERK/MAPK pathways. Indeed, constitutively active PAK mutants can activate both the SAPK and p38 pathways in many cell types [45,53,54]. However, mutant Rac1 and Cdc42Hs constructs which are GTPase-deficient, but cannot bind PAKs (C40/L61), are still perfectly capable of activating co-expressed SAPK [55,56]. Moreover, at least two MAPKKKs upstream of the SAPKs, MEKK4 and SPRK/MLK3, contain CRIB motifs [30,50], and may be the Rac1 and Cdc42Hs targets which couple to the SAPKs and p38s. Thus a role for the PAKs in SAPK and p38 regulation cannot be assigned unambiguously at this time.

Mammalian homologues of SPS1: activators of stress signalling pathways

The SAPK pathway can also be regulated, independently of Rac1 and Cdc42Hs, through protein kinases of the SPS1 family. Sps1p is a *S. cerevisiae* Ser/Thr kinase required for the encapsulation of haploid nuclei during sporulation [57]. It consists of an N-terminal kinase domain and an extensive C-terminal domain with no obvious structural features [57]. Six mammalian SPS1 homologues have been reported thus far [58–64]. All mammalian SPS1 homologues appear to regulate preferentially stress signalling pathways; however, whereas some of these kinases can activate the SAPKs *in situ* [63–65], no mammalian SPS1 homologues have yet been identified as upstream activators of the p38s. The kinase domains of SPS1s are distantly related to that of Ste20p; thus the SPS1s were originally classified as STE20 homologues [58–64]. However, as the SPS1 family enlarges with the cloning and characterization of additional mammalian kinases, it has become clear that the SPS1 family is a distinct protein kinase subgroup with unique regulatory and functional properties. Figure 3 illustrates the structural features of the three human SPS1 family kinases implicated in SAPK regulation.

Germinal centre kinase (GCK) was the first mammalian SPS1 homologue to be identified. The enzyme was cloned during a subtractive screen for cDNAs selectively expressed in the B follicular germinal centre and not in the surrounding mantle zone [58]. B-lymphocyte selection occurs in the germinal centre, and this site is a region of extensive apoptosis and B-cell maturation; these processes are mediated in part by

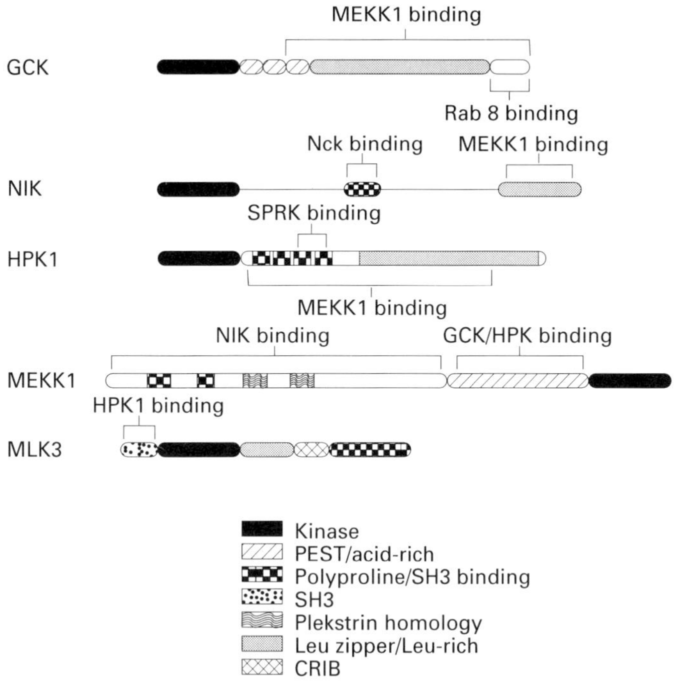

Fig. 3. Schematic illustration of the domain structures of mammalian SPS1s and their known MAPKKK targets. The binding domains are highlighted.

receptors of the TNF family, including TNFR1, Fas and CD40. Because of the association between the germinal centre and B-cell selection/ maturation, as well as the activation of the SAPKs by TNF, we reasoned that GCK might activate the SAPKs. Indeed, overexpression of GCK potently activates the SAPKs. Moreover, this activation is quite selective; overexpression of GCK fails to activate the p38 or MAPK pathways [65].

GCK activation of the SAPKs occurs in the absence of external ligand, and the kinase is enzymically constitutively active when overexpressed [58,65]. This contrasts with the PAKs, which must be activated by mutation or co-expression with active forms of Rac1 or Cdc42 [45,51–54]. This finding suggested that GCK is activated either by dissociation of limiting concentrations of an inhibitor or by oligomerization, both processes which would be promoted by GCK overexpression. The GCK C-terminal non-catalytic domain consists of three PEST motifs followed by a leucine-rich motif and a short Rab8 binding domain (Fig. 3; and see below) [58]. The GCK C-terminal domain is probably the site of GCK regulation and effector coupling. In support of this, overexpression of the GCK C-terminal domain results in substantial activation of co-expressed SAPK.

In addition to GCK, at least two other mammalian *SPS1* homologues activate the SAPK pathway: haematopoietic progenitor kinase-1 (HPK1) and Nck-interacting kinase (NIK) [62–64]. Like GCK, both HPK1 and NIK are selective for the SAPK pathway and fail to activate co-expressed MAPK or p38. In addition, as with GCK, overexpression of the NIK C-terminal domain results in modest activation of co-expressed SAPK [64]. By contrast, overexpression of the HPK C-terminal domain does not activate co-expressed SAPK [62,63].

Three additional mammalian SPS1 homologues have been cloned: kinase responsive to stress-1 (Krs1), mammalian sterile twenty-like-1 (MST1)/Krs2 and Ste20-like oxidant-stress-activated kinase-1 (SOK1) [59–61]. Upon overexpression, none of these SPS1 homologues activates any of the known ERK/MAPK pathways [59,61]. These kinases possess significant basal activity when immunoprecipitated from endogenous sources, or when overexpressed. However, in contrast with HPK1, NIK and GCK, which are largely refractory to further activation by extracellular agonists, Krs1, MST1/Krs2 and SOK1 can be activated significantly *in situ* by different environmental stresses [58–64]. Krs1 and MST1/Krs2 are activated by extreme heat shock and by high concentrations of arsenite, staurosporine and okadaic acid. MST1/Krs2 can also be activated *in vitro* by protein phosphatase-2A [52,59,61]. Thus these two SPS1-like kinases may be activated by both phosphorylation and dephosphorylation. SOK1, as its name implies, is strongly activated by oxidative stress [60]. SOK1 is also activated by ischaemic injury and depletion of the cellular ATP pool. In all cases, SOK1 activation appears to require the generation of reactive oxygen intermediates as well as elevated levels of cytosolic free calcium (J. Bonventre, T. Force and J.M. Kyriakis, unpublished work).

The SPS1s target MAPKKKs directly through physical association

The first hints as to the mechanism of action of mammalian SPS1 homologues came with the finding that these kinases can bind MAPKKKs (Figs. 2 and 3). Thus HPK1 can bind both MEKK1 and SPRK/MLK3 [62,63]. These interactions require the HPK1 C-terminal domain. The HPK1–MEKK1 interaction requires the C-terminal half (residues 817–1221) of the MEKK1 regulatory domain, whereas the HPK1–SPRK/MLK3 interaction has been mapped to the C-terminal two of four polyproline stretches in the HPK1 C-terminal domain which interact with the SPRK/MLK3 SH3 domain (Fig. 3) [62,63]. Both SPRK/MLK3 and MEKK1 can bind full-length HPK1 and the isolated HPK1 C-terminal tail with apparently similar affinities; thus it is not clear how the HPK1–MEKK1 and the HPK1–SPRK/MLK3 interactions are regulated [62,63]. Nor is it clear if MEKK1 and SPRK/MLK3 can compete for binding to HPK1 in response to common upstream stimuli, or if pre-existing complexes of HPK1 and MEKK1 or HPK1 and SPRK/MLK3 respond to different stimuli. Although an interaction between expressed HPK1 and endogenous MEKK1 or SPRK/MLK3 has not been detected, it is possible that both of these MAPKKKs are HPK1 targets. Consistent with this idea is the observation that kinase-inactive mutants of MEKK1 and SPRK/MLK3 can effectively block HPK1 activation of the SAPKs [62,63].

NIK was cloned in a two-hybrid screen which sought to identify polypeptides that could bind the SH2/SH3 adaptor Nck (Figs. 2 and 3) [64]. The NIK C-terminal regulatory domain contains a single polyproline SH3 binding domain which mediates the interaction with Nck. It is not known if the Nck–NIK interaction couples NIK to tyrosine kinases in a manner analogous to the coupling of SOS to receptor tyrosine kinases by Grb2 [64]. The C-terminal 326 amino acids of NIK are necessary for the observed interaction between NIK and MEKK1. This domain interacts with the N-terminal 719 amino acids of MEKK1. The remainder of the MEKK1 polypeptide is not necessary for the MEKK1–NIK interaction [64].

The C-terminal MEKK1 binding site of NIK forms a hydrophobic motif which is analogous to the GCK Leu-rich domain (amino acids 463–819) and part of the C-terminal extension of HPK1 (amino acids 470–830) [58,62–64]. The fact that this region has been conserved among the three SPS1 family kinases that can bind MEKK1 has led to the idea that this domain might form a common MEKK1 binding site [64]. Our results with a truncated MEKK1 clone indicate that the situation is more complicated (see below). Moreover, while the conserved C-terminal domain of NIK could form a contact point with the N-terminal 719 amino acids of MEKK1, HPK1 and GCK can bind an MEKK1 construct wherein residues 1–719 are deleted (Fig. 3) (see below and [63]). Whether or not the Leu-rich portion of the HPK1 C-terminal domain mediates the HPK–MEKK1 interaction has not been determined. As with the HPK–MEKK1 interaction, kinase-inactive MEKK1 can inhibit NIK activation of the SAPK pathway, suggesting that MEKK1 is a physiological target of NIK [64].

We have observed an interaction between GCK and either endogenous or co-expressed, recombinant MEKK1. In contrast with the results with NIK and HPK, we observe that GCK C-terminal domain interacts with MEKK1 much more stably than does full-length GCK, and that kinase-inactive GCK barely interacts with MEKK1 at all. Mapping studies indicate that the interaction between MEKK1 and GCK requires the C-terminal domain of GCK, but does not require an intact Leu-rich motif. Truncation studies with the MEKK1 polypeptide reveal that MEKK1 binds GCK through an acid-rich domain (amino acids 817–1221) which is clearly distinct from the domain of MEKK1 that binds NIK [64]. Expression of this GCK binding domain of MEKK1 effectively blocks GCK activation of co-expressed SAPK, indicating that MEKK1 is a true GCK target.

Regulation of SPS1 family kinases by Rab family GTPases

Recently, a two-hybrid screen employing Rab8 as a bait demonstrated that GCK could bind Rab8 *in situ* [66]. We have extended these results, demonstrating that the interaction can be reproduced *in vitro* and that binding is GTP-dependent. We have mapped the Rab8 binding site on GCK to the C-terminal 139 amino acids of the GCK polypeptide. Rab family GTPases are members of the Ras superfamily and have been implicated in the regulation of vesicular trafficking. Rab8 regulates traffic between the *trans*-Golgi and the plasma membrane [67]. Thus GCK may play an effector role in Rab8-regulated vesicle movement or, alternatively, Rab8 may target GCK to substrate proteins associated with the cytosolic leaflets of vesicle membranes destined for fusion with the plasma membrane. We do not see activation of SAPK upon co-expression with GTPase-deficient Rab8 mutants; thus Rab8 does not activate signalling machinery upstream of the SAPKs in a manner analogous to the action of Ras, Rac1 or Cdc42Hs.

Regulation of the SAPK pathway by SPS1 family kinases

The findings described above indicate that SPS1 family kinases can interact with MAPKKKs (MEKK1 and SPRK/MLK3) constitutively upon overexpression. These interactions are consistent with the ability of these kinases to activate the SAPK pathway in the absence of extracellular stimuli [62–65]. How might these interactions be regulated *in vivo*? The results with GCK are somewhat informative in this matter. Full-length GCK interacts poorly with MEKK1, whereas the GCK C-terminal domain forms a stable complex with MEKK1. Kinase-inactive GCK interacts with MEKK1 even more poorly than does wild-type GCK. By contrast, full-length, wild-type GCK is a more potent activator of co-expressed SAPK than is the GCK C-terminus or kinase-inactive GCK. The simplest reconciliation of these apparently conflicting results is to propose that activation of GCK's kinase activity allows MEKK1 binding and provides for an efficient recruitment and release of activated MEKK1. It is not known if MEKK1 activation involves GCK-catalysed phosphorylation; however, inasmuch as the GCK C-terminus and kinase-inactive GCK can both appreciably activate the SAPK pathway *in situ* [65], direct phosphorylation of MEKK1 by GCK may not be necessary for activation of the SAPK pathway. An equally plausible scenario is that GCK and other SPS1 kinases bind MAPKKKs and bring them to sites of activation mediated by separate components, or to sites where substrates are located. Such a function for SPS1 kinases would be that of a regulated scaffolding protein.

The regulation of SPS1 kinases by upstream stimuli is also unclear. It remains to be determined if Rab family GTPases regulate the binding of MAPKKKs. All SPS1 family kinases display significant basal activity [58–65]; however, some members of the SPS1 family can be immunoprecipitated from cells in either a 'resting' or agonist-activated state (Krs1, MST/Krs2, SOK1), while others (GCK) display a basal activity so high as to mask potential activation by most extracellular stimuli [58–65]. This may be due to technical problems, such as inappropriate isolation/assay conditions. Alternatively, the high basal activity may be due to autoactivation of the kinase by the aggregation artificially induced during formation of the immune complex. The mechanism of activation of SPS1 family kinases will be the focus of much future work.

TNF receptor (TNFR) signalling and the SAPKs

The SAPKs and p38s are potently activated by inflammatory cytokines of the TNF family [1,3,4]. The TNFR family includes the two TNFRs (the 55 kDa TNFR1/CD120a and the 75 kDa TNFR2/CD120b), as well as Fas, the IL-1 receptor, CD40, CD30 and CD27 [68]. TNFR family members share extensive identity within their extracellular domains, but differ widely in their intracellular extensions [68]. TNF itself is a model ligand for this group, and recent studies have elucidated some of the molecular players that couple TNF to the SAPKs.

TNF is a homotrimeric polypeptide, and TNF binding to TNFRs promotes receptor trimerization and initiates signal transduction [68,69]. A significant advancement in the understanding of TNF signalling came with the identification of polypeptide species that are recruited to the TNFRs as a consequence of ligand-induced receptor trimerization. Many of these proteins, upon overexpression, mimic the cellular responses to TNF. The intracellular domains of TNFR1 and Fas, but not those of TNFR2, CD40 or the IL-1 receptor, contain an ~80-amino-acid motif termed the 'death domain', which is critical for signalling apoptosis [68,69]. The death domain mediates both homotypic and heterotypic protein–protein interactions. Thus the TNFR1 death domain binds a protein termed TNFR-associated death domain protein (TRADD) [70]. TRADD, in turn, binds a related species, Fas-associated death domain protein (FADD) [71]. FADD can also interact directly with the death domain of Fas. FADD is necessary for the apoptotic responses to both TNF and Fas.

TRADD can also bind TNFR-associated factor 2 (TRAF2) [71]. TRAF2 is one of an emerging family of signal transducers which consist of a C-terminal TRAF domain, necessary for interaction with upstream and downstream signalling components, and (with the exception of TRAF1) an N-terminal RING (really interesting new gene) finger domain which probably mediates the activation of downstream signalling elements [72]. The TRAF domain of TRAF2 interacts with an N-terminal TRAF-interacting domain on the TRADD polypeptide, thereby coupling TRAF2 to TNFR1 [71]. By contrast, TNFR2 can bind TRAF2 directly, a process also mediated by the TRAF domain [72]. TRAF2 appears to be necessary for the induction of nuclear factor κB (NF-κB) by both TNFR1 and TNFR2 [72].

TRAF2 and the SAPKs

Recently it was reported that overexpression of TRAF2 could activate co-expressed SAPK, and that expression of a mutant TRAF construct [TRAF2-(87–501)] in which the RING finger was deleted blocked TNF activation of the SAPKs (Fig. 2) [73,74]. This construct probably exerted its effects by binding TRADD and sequestering it from endogenous TRAF2 [71]. These results suggest that TRAF2 is necessary for coupling TNFR1 to the SAPKs. Interestingly, TRADD, which is necessary for recruiting TRAF2 to TNFR1 [70], failed to activate co-expressed SAPK [73,74]. Thus it is possible that as yet unidentified polypeptides can bind TRAF2 to TNFR1 and mediate SAPK activation.

Receptor-interacting protein (RIP) and the SAPKs

RIP was originally identified as a polypeptide that interacts with the death domain of Fas [75]. RIP consists of an N-terminal protein Ser/Thr kinase domain (amino acids 1–304), a C-terminal death domain (amino acids 554–671) and an intermediate domain (amino acids 305–553) [75]. Expression of RIP was found to recapitulate many of the signalling events associated with Fas or TNFR1 engagement. Thus overexpression of RIP promotes apoptosis and activates NF-κB [75,76]. Nevertheless, in spite of the observed interaction between Fas and RIP, it has become clear that RIP is a major component of TNFR1, but not Fas, signalling [76,77]. Notably, mutant Jurkat cells devoid of RIP expression display a complete loss of TNF-induced NF-κB induction, whereas Fas-induced apoptosis is unaffected [76].

The RIP death domain interacts strongly with that of TRADD, and TNF treatment recruits the TRADD–RIP complex to TNFR1 [77]. RIP can also bind TRAF2. This interaction involves the TRAF domain of TRAF2 and both the kinase and intermediate domains of RIP [77]. The recruitment of TRAF, TRADD and RIP to TNFR1 is accordingly complex. It appears that TRADD can recruit TRAF2 and RIP to TNFR1 simultaneously via TRADD's N-terminal TRAF-interacting domain and

C-terminal death domain respectively. Within the receptor complex, TRAF2 and RIP presumably interact via the RIP intermediate and kinase domains and the TRAF domain of TRAF2 [77].

Deletion experiments have identified the RIP intermediate domain as being required for NF-κB induction [76]. Thus kinase-inactive RIP mutants, when overexpressed, are capable of activating NF-κB constitutively. By contrast, deletion of the intermediate domain eliminates NF-κB induction. Further studies have identified a charged motif within the RIP intermediate domain (amino acids 391–422) which is absolutely necessary for NF-κB induction. Deletion of this segment results in a RIP mutant that is incapable of inducing NF-κB [76].

Overexpression of RIP also activates the SAPK pathway (Fig. 2) [73]. While activation of the SAPKs by GCK is partially abrogated upon truncation of the kinase domain, RIP kinase activity is completely unnecessary for SAPK and p38 recruitment by RIP [73]. Expression of the RIP death domain is sufficient to block TNF activation of the SAPK pathway. This result is analogous to inhibition of SAPK activation by TRAF2-(87–501), and may merely reflect sequestration of TRADD and TNFR1 from *in vivo* SAPK activation machinery. Alternatively, both RIP and TRAF2 may be necessary for SAPK activation. Whether or not TRAF2 signals through RIP or vice versa is not known.

As is the case with NF-κB activation, the RIP intermediate domain appears to be necessary for SAPK and p38 activation by RIP, inasmuch as the RIP death domain fails to activate co-expressed SAPK, whereas a construct consisting of the intermediate domain and the death domain strongly activates co-expressed SAPK [73].

It is not known how TRAF2 or RIP recruits SAPK or p38 core signalling modules. It has been shown that TNF can induce the production of ceramide via activation of sphingomyelinases, and addition of exogenous ceramide or sphingomyelinase can activate SAPK [1,3,4,78]. Thus it is possible that TRAF2 and RIP couple TNFR1 to a sphingomyelinase. The role of ceramide in SAPK activation by TNF is not clear, however. While TNF does stimulate ceramide generation, and exogenous ceramide can activate the SAPKs, the kinetics of TNF-induced ceramide generation do not always coincide with those of TNF activation of the SAPKs [1,3,4,78,79].

It is equally possible that TRAF2 and RIP couple directly to MAPKKKs. A precedent for this possibility comes from the identification of a novel MAPKKK, NF-κB-inducing kinase (NIK; to avoid confusion with the SPS1 family kinase Nck-interacting kinase described above, I will refer to the MAPKKK NIK as $\mathrm{NIK_{MAPKKK}}$) [80]. The kinase domain of $\mathrm{NIK_{MAPKKK}}$ bears significant sequence identity with those of MEKK1–MEKK4 as well as with those of other kinases similar in structure to *S. cerevisiae STE11*. $\mathrm{NIK_{MAPKKK}}$ associates directly with TRAF2 and appears to be necessary for activation of NF-κB by TNF, Fas and IL-1 [80]. It remains to be determined whether or not $\mathrm{NIK_{MAPKKK}}$ can also activate the SAPK or p38 pathways. RIP does not associate with $\mathrm{NIK_{MAPKKK}}$ [80]; thus

NIK_{MAPKKK} probably does not couple RIP to the SAPKs. While it remains to be determined if RIP associates with MAPKKKs upstream of the SAPKs and p38s, the identification of NIK_{MAPKKK} makes such a mechanism an attractive hypothesis. Should such a hypothesis prove to be correct, then TRAF2 and RIP would perform a scaffolding/recruitment role similar to that of GCK and the SPS1 family kinases.

Role of the SAPKs and p38s in the regulation of cell cycle progression and apoptosis

The biological functions of the SAPKs and p38s are only beginning to be identified. The SAPKs and p38s are regulated by extracellular stimuli which are known to arrest the cell cycle (ionizing radiation, heat shock, chemical DNA damage) or promote apoptosis (TNF, in addition to the above) [1,3,4]. The role of the SAPKs and p38s in these processes has been the focus of much study and controversy. What follows is a brief summary of these often conflicting results.

A number of reports have implicated the Rho family GTPases Rac1 and Cdc42Hs in cell-cycle entry [47,48]. Most of these reports have utilized Swiss 3T3 cells or Rat1 cells. Specifically, it has been shown that blockade of Cdc42Hs or Rac1 function by overexpression of Asn^{17} dominant-inhibitory mutants prevents G0 cells from transiting through G1/S [48]. In addition, [Asn^{17}]Rac1 significantly inhibits the transforming potential of stably expressed oncogenic *ras* mutants, suggesting that Ras regulates Rac1 [47]. Indeed, it has been shown that Rac1-activated changes in the cortical actin cytoskeleton, such as membrane ruffling, are essential processes in the Ras signalling programme [46], and that activation of SAPK by oncogenic Ras or epidermal growth factor requires Rac1 [41].

Our results with Cdc42Hs are quite different. We have microinjected normal, serum-stimulated NIH 3T3 cells in early G1 with components of the p38 signalling pathway and monitored the cells' progression through G1/S. Expression of wild-type p38, MKK3 and MKK6 could effectively inhibit serum-stimulated G1/S progression [49]. SAPK was without significant effect. These results are consistent with reports suggesting that mitogen induction of cyclin D1 expression can be inhibited by p38 [81]. Moreover, expression of wild-type Cdc42Hs could also effectively block G1/S progression through a process that could be partially inhibited by kinase-inactive MKK3, indicating a role for p38 in Cdc42Hs-mediated cell cycle arrest [49]. Interestingly, Rac1 was just as effective as Cdc42Hs at activating p38, but was unable to promote cell cycle arrest [49]. Thus Rac1 may indeed be an essential component of the Ras-stimulated growth of NIH 3T3 cells. By contrast, Cdc42Hs clearly is not; it is, in fact, growth inhibitory for NIH 3T3 cells. Moreover, given the equivalent ability of Rac1 and Cdc42Hs to activate p38 [49], our results suggest that additional signalling pathways emanating from these G-proteins may mediate cell cycle inhibition in the case of Cdc42Hs or cell cycle progression in the case of Rac1.

Whereas our results concern cells already in G1, the results in Swiss 3T3 and Rat1A cells pertain to cell-cycle entry, and may reflect the combined effects of Ras and Cdc42. Alternatively, use of dominant-inhibitory Cdc42Hs could sequester GEFs shared with Rac. In support of this idea, we observed that [Asn^{17}]Cdc42Hs inhibits serum-stimulated cell cycle progression without activating p38 [49]; in this instance the mutant Cdc42Hs could, by sequestering GEFs upstream of Rac, prevent the activation of Rac1 which is necessary for growth.

Studies of the roles of the SAPKs and p38s in apoptosis have been equally confounding, and it is reasonable to conclude that their roles are cell- and stimulus-specific. In particular, it is likely that the SAPKs and p38s mediate apoptotic mechanisms that require new gene expression. Initial reports suggesting a role for the SAPKs in apoptosis came from studies of PC12 cells. When these cells are serum-starved, a modest proportion of them will undergo apoptosis. Expression of dominant inhibitors of SAPK or p38 activation (kinase-inactive SEK1 or MKK3 respectively) prevents much of this apoptosis [82]. Constitutively active forms of SEK1 or MKK3 promote apoptosis in the presence of serum, and activation of the MAPK pathway with constitutively active MEK prevents apoptosis upon serum withdrawal [82]. This last result suggests that a dynamic balance between MAPK and SAPK/p38 activation influences the decision to undergo apoptosis [82].

Zanke et al. [83] characterized a thermotolerant fibroblast line which displays little or no heat-shock activation of SAPK as compared with the parental cell line. The parental cell line can be stimulated to undergo apoptosis upon heat stress, whereas the mutant cell cannot. Blockade of SAPK activation by overexpression of a kinase-inactive SEK1 mutant renders the parental cells thermotolerant, indicating that heat-shock-induced cell death is SAPK-dependent in these cells [83].

Ceramide is a potent inducer of apoptosis in U937 cells [78]. Verheij et al. [78] demonstrated that TNF, as well as a variety of environmental stresses, results in the generation of ceramide from membrane sphingomyelin. Inhibition of ceramide- and TNF-induced apoptosis can be achieved upon stable expression of a dominant-negative c-Jun construct, TAM67, wherein the sites of SAPK phosphorylation have been deleted [78]. In addition, stable expression of a kinase-inactive, dominant-inhibitory SEK1 construct also prevents ceramide- and TNF-induced apoptosis. The results with the SEK-KR and TAM67 constructs support the contention that, in cases where ceramide couples TNF and stress to the SAPKs, the SAPKs are important in apoptotic mechanisms that involve *de novo* gene expression [78].

Contrasting results have been obtained in studies of MCF7 breast cancer cells, in which TNF-induced apoptosis appears not to require the SAPK pathway. Thus, in these cells, overexpression of a kinase-inactive

MEKK1 catalytic domain construct, which probably acts to sequester all MEKK1 substrates, including SEK1, fails to protect MCF7 cells from TNF-induced apoptosis [73].

Studies of Fas-induced apoptosis have uncovered widely divergent results concerning the role of the SAPKs and p38s in apoptosis. Targeted disruption of the *SEK1* gene proved embryonically lethal; however, it was possible to generate chimaeric mice using the SEK knockout embryonic stem cells to reconstitute the immune system of RAG (recombination-activating gene) 1-deficient embryos [20]. Accordingly, these mice bear an SEK1-deficient immune system. It was observed in this system that the loss of SEK1 made T-lymphocytes more sensitive to Fas-induced apoptosis. It was concluded from this result that SEK1 protects T-cells from Fas-induced apoptosis [20].

In Jurkat cells, Fas-induced apoptosis requires caspase family pro-apoptotic proteases and MKK6 [84]. MKK6 activation appears to require caspases and, in turn, promotes the activation/induction of other caspases [84]. Interestingly, in spite of the requirement for MKK6, expression of dominant-inhibitory constructs of p38 failed to block Fas-induced apoptosis, indicating that p38 is not necessary for Fas-induced apoptosis, and that MKK6 may recruit an as yet unidentified substrate as part of its apoptotic programme.

While there have been reports indicating that Fas-induced apoptosis in Jurkat cells is not blocked by the TAM67 construct, and therefore probably does not require the SAPK pathway [85], at least two studies have implicated the SAPKs in Fas-induced apoptosis. SHEP cells are a neuroblastoma cell line which undergoes rapid Fas-induced apoptosis. Stable expression of a dominant interfering SAPK construct wherein the sites of SEK phosphorylation, Thr-183 and Tyr-185, have been mutated to Ala and Phe respectively can partially inhibit Fas-induced apoptosis [86].

In addition, an elegant study by Yang et al. [87] identified a second, novel, Fas-induced pathway which promotes apoptosis in collaboration with the FADD pro-apoptotic pathway. Daxx is a polypeptide which binds to the Fas death domain and mediates Fas activation of the SAPKs. Daxx overexpression promotes apoptosis through a process which involves the induction of caspases. Overexpression of TAM67 or dominant-inhibitory SEK1 effectively blocks Fas- and Daxx-induced apoptosis in L929 and 293 cells (but not in HeLa cells) without interfering with FADD-induced apoptosis [87]. These results clearly show a cell-specific requirement for the SAPKs in Fas-induced apoptosis that is SAPK-dependent.

Conclusions

This review has focused on the recruitment of mammalian stress-activated ERK/MAPK pathway core modules by upstream kinases of the SPS1 family and by adaptor proteins recruited by activated TNFR1. I have also touched on some recent results concerning the biological consequences of SAPK and p38 activation. It is evident that the diversity of MAPKKKs which feed into the SAPKs and p38s allows for these kinase modules to interact with a wide range of upstream components, including small GTPases of the Rho subfamily, SPS1 family kinases and adaptor proteins coupled to cytokine receptors (Fig. 3). In the case of the SPS1-like kinases, it remains to be determined which upstream ligands activate them and how they are activated. It is intriguing to speculate that these kinases may themselves bind adaptor proteins such as TRAF2 or Daxx. Alternatively, these kinases may associate directly with receptors.

The mode of activation of SPS1 family kinases and RIP is also unclear. In particular, inasmuch as the kinase domains of these proteins appear to be largely dispensable, what is the role of SPS1-family or RIP kinase activity? As mentioned above, the kinase activity of GCK could promote rapid, more efficient, recruitment of MEKK1. The kinase activity of RIP could also, once activated, promote the efficient recruitment of effectors. In support of this, we observe that RIP constructs with intact intermediate and kinase domains are expressed less well than is the free intermediate domain, yet activate the SAPK pathway equally well. Thus RIP constructs with active kinase domains possess a higher 'specific activity' with regard to recruitment of the SAPKs and p38s. Alternatively, the kinase domains of RIP and GCK could couple to downstream components distinct from the SAPKs.

The biological functions of the SAPKs and p38s are clearly quite varied and often conflicting. At this point, it is safe to say that the roles of these kinases reflect the roles in the stress response (growth, death, repair, etc.) played by the cells in which the kinases operate, and the magnitude of the stressful stimulus. Investigators should be cautious in this regard about drawing encyclopaedic conclusions as to the functions of these kinases based on results from a single cell type. There is clearly much more exciting work to be done.

I thank members of my laboratory, notably Celia Pombo, Takashi Yuasa, Árpád Molnár and Anthony Makkinje, for their tireless efforts which produced many of the results described herein. I also thank James Woodgett, Joseph Avruch, John Kehrl, Brian Seed and Thomas Force for valuable collaborations and discussions.

References

1. Kyriakis, J.M. and Avruch, J. (1996) J. Biol. Chem. **271**, 24313–24316
2. Herskowitz, I. (1995) Cell **80**, 187–197
3. Kyriakis, J.M. and Avruch, J. (1996) Bioessays **18**, 567–577
4. Kyriakis, J.M., Banerjee, P., Nikolakaki, E., Dai, T., Rubie, E.A., Ahmad, M.F., Avruch, J. and Woodgett, J.R. (1994) Nature (London) **369**, 156–160
5. Dérijard, B., Hibi, M., Wu, I.-H., Barrett, T., Su, B., Deng, T., Karin, M. and Davis, R.J. (1994) Cell **76**, 1025–1037
6. Han, J., Lee, J.-D., Bibbs, L. and Ulevitch, R.J. (1994) Science **265**, 808–811
7. Rouse, J., Cohen, P., Trigon, S., Morange, M., Alonso-Llamazares, A., Zamanillo, D., Hunt, T. and Nebreda, A. (1994) Cell **78**, 1027–1037

8. Freshney, N.W., Rawlinson, L., Guesdon, F., Jones, E., Cowley, S., Hsuan, J. and Saklatvala, J. (1994) Cell **78**, 1039–1049
9. Jiang, Y., Chen, C., Li, Z., Guo, W., Gegner, J.A., Lin, S. and Han, J. (1996) J. Biol. Chem. **271**, 17920–17926
10. Gupta, S., Barrett, T., Whitmarsh, A.J., Cavanagh, J., Sluss, H., Dérijard, B. and Davis, R.J. (1996) EMBO J. **15**, 2760–2770
11. Karin, M., Liu, Z.-G. and Zandi, E. (1997) Curr. Opin. Cell Biol. **9**, 240–246
12. Brenner, D.A., O'Hara, M., Angel, P., Chojkier, M. and Karin, M. (1989) Nature (London) **337**, 661–663
13. Devary, Y., Gottleib, R.A., Smeal, T. and Karin, M. (1992) Cell **71**, 1081–1091
14. Pulverer, B.J., Kyriakis, J.M., Avruch, J., Nikolakaki, E. and Woodgett, J.R. (1991) Nature (London) **353**, 670–674
15. Mertens, S., Craxton, M. and Goedert, M. (1996) FEBS Lett. **383**, 273–276
16. Goedert, M., Cuenda, A., Craxton, M., Jakes, R. and Cohen, P. (1997) EMBO J. **16**, 3563–3571
17. Han, J., Jiang, Y., Li, Z., Kravchenko, V.V. and Ulevitch, R.J. (1997) Nature (London) **386**, 563–566
18. Sánchez, I., Hughes, R.T., Mayer, B.J., Yee, K., Woodgett, J.R., Avruch, J., Kyriakis, J.M. and Zon, L.I. (1994) Nature (London) **372**, 794–798
19. Dérijard, B., Raingeaud, J., Barrett, T., Wu, L.-H., Han, J., Ulevitch, R.J. and Davis, R.J. (1995) Science **267**, 682–685
20. Nishina, N., Fischer, K.D., Radvanyl, L., Shahinlan, A., Hakem, R., Rubie, E.A., Bernstein, A., Mak, T.W., Woodgett, J.R. and Penninger, J.M. (1997) Nature (London) **385**, 350–353
21. Mizukami, Y., Yoshioka, K., Morimoto, S. and Yoshida, K.-i. (1997) J. Biol. Chem. **272**, 16657–16662
22. Meier, R., Rouse, J., Cuenda, A., Nebreda, A. and Cohen, P. (1996) Eur. J. Biochem. **236**, 796–805
23. Raingeaud, J., Whitmarsh, A.J., Barrett, T., Dérijard, B. and Davis, R.J. (1996) Mol. Cell. Biol. **16**, 1247–1255
24. Cuenda, A., Cohen, P., Buée-Scherrer, V. and Goedert, M. (1997) EMBO J. **16**, 295–305
25. Moriguchi, T., Kuroyanagi, N., Yamaguchi, K., Gotoh, Y., Irie, K., Kano, T., Shirakabe, K., Muro, Y., Shibuya, H., Matsumoto, K., et al. (1996) J. Biol. Chem. **271**, 13675–13679
26. Lange-Carter, C.A., Pleiman, C.M., Gardner, A.M., Blumer, K.J. and Johnson, G.L. (1993) Science **260**, 315–319
27. Yan, M., Dai, T., Deak, J.C., Kyriakis, J.M., Zon, L.I., Woodgett, J.R. and Templeton, D.J. (1994) Nature (London) **372**, 798–800
28. Blank, J.L., Gerwins, P., Elliot, E.M., Sather, S. and Johnson, G.L. (1996) J. Biol. Chem. **271**, 5361–5368
29. Ellinger-Ziegelbauer, H., Brown, K., Kelly, K. and Siebenlist, U. (1997) J. Biol. Chem. **272**, 2668–2674
30. Gerwins, P., Blank, J.L. and Johnson, G.L. (1997) J. Biol. Chem. **272**, 8288–8295
31. Takekawa, M., Posas, F. and Saito, H. (1997) EMBO J. **16**, 4973–4982
32. Ichijo, H., Nishida, E., Irie, K., ten Dijke, P., Saitoh, M., Moriguchi, T., Takagi, M., Matsumoto, K., Miyazono, K. and Gotoh, Y. (1997) Science **275**, 90–94
33. Rana, A., Gallo, K., Godowski, P., Hirai, S.-I., Ohno, S., Zon, L.I., Kyriakis, J.M. and Avruch, J. (1996) J. Biol. Chem. **271**, 19025–19028
34. Hirai, S.-i., Katoh, M., Terada, M., Kyriakis, J.M., Zon, L.I., Rana, A., Avruch, J. and Ohno, S. (1997) J. Biol. Chem. **272**, 15167–15173

35. Yamaguchi, K., Shirakabi, K., Shibuya, H., Irie, K., Oishi, I., Ueno, N., Taniguchi, T., Nishida, E. and Matsumoto, K. (1995) Science **270**, 2008–2011
36. Salmerón, A., Ahmad, T.B., Carlile, G.W., Pappin, D., Narsimhan, R.P. and Ley, S.C. (1996) EMBO J. **15**, 817–826
37. Zanke, B.W., Rubie, E.A., Winnett, E., Chan, J., Randall, S., Parsons, M., Boudreau, K., McInnis, M., Yan, M., Templeton, D.J. and Woodgett, J.R. (1996) J. Biol. Chem. **271**, 29876–29881
8. Deacon, K. and Blank, J.L. (1997) J. Biol. Chem. **272**, 14489–14496
39. Marshall, C.J. (1995) Cell **80**, 179–185
40. Avruch, J., Zhang, X.-f. and Kyriakis, J.M. (1994) Trends Biochem. Sci. **19**, 279–283
41. Ridley, A.J., Paterson, H.F., Johnston, C.L., Diekmann, D. and Hall, A. (1992) Cell **70**, 401–410
42. Nobes, C.D. and Hall, A. (1995) Cell **81**, 53–62
43. Coso, O.A., Chiarello, M., Yu, J.-C., Teramoto, H., Crespo, P., Xu, N., Miki, T. and Gutkind, J.S. (1995) Cell **81**, 1137–1146
44. Minden, A., Lin, A., Claret, F.-X., Abo, A. and Karin, M. (1995) Cell **81**, 1147–1157
45. Zhang, S., Han, J., Sells, M.A., Chernoff, J., Knaus, U.G., Ulevitch, R.J. and Bokoch, G.M. (1995) J. Biol. Chem. **270**, 23934–23936
46. Joneson, T., McDonough, M., Bar-Sagi, D. and Van Aelst, L. (1996) Science **274**, 1374–1376
47. Qiu, R.-G., Chen, J., Kirn, D., McCormick, F. and Symons, M. (1995). Nature (London) **374**, 457–459
48. Olson, M.F., Ashworth, A. and Hall, A. (1995) Science **269**, 1270–1272
49. Molnár, Á., Theodoras, A.M., Zon, L.I. and Kyriakis, J.M. (1997) J. Biol. Chem. **272**, 13229–13235
50. Burbelo, P.D., Drechsel, D. and Hall, A. (1995) J. Biol. Chem. **270**, 29071–29074
51. Manser, E., Leung, T., Salihuddin, H., Zhao, Z.-S. and Lim, L. (1994) Nature (London) **367**, 40–46
52. Sells, M.A. and Chernoff, J. (1997) Trends Cell Biol. **7**, 162–167
53. Bagrodia, S., Dérijard, B., Davis, R.J. and Cerione, R.A. (1995) J. Biol. Chem. **270**, 27995–27998
54. Polverino, A., Frost, J., Yang, P., Hutchinson, M., Neiman, A.M., Cobb, M.H. and Marcus, S. (1995) J. Biol. Chem. **270**, 26067–26070
55. Lamarche, N., Tapon, N., Stowers, L., Burbelo, P., Aspenström, P., Bridges, T., Chant, J. and Hall, A. (1996) Cell **87**, 519–529
56. Westwick, J.K., Lambert, Q.T., Clark, G.J., Symons, M., Van Aelst, L., Pestell, R.G. and Der, C.J. (1997) Mol. Cell. Biol. **17**, 1324–1335
57. Friesen, H., Lunz, R., Doyle, S. and Segall. J. (1994) Genes Dev. **8**, 2162–2175
58. Katz, P., Whalen, G. and Kehrl, J.H. (1994) J. Biol. Chem. **269**, 16802–16809
59. Creasy, C.L. and Chernoff, J. (1995) J. Biol. Chem. **270**, 21695–21700
60. Pombo, C.M., Bonventre, J.V., Molnár, A., Kyriakis, J. and Force, T. (1996) EMBO J. **15**, 4537–4546
61. Taylor, L.K., Wang, H.-C.R. and Erikson, R.L. (1996) Proc. Natl. Acad. Sci. U.S.A. **93**, 10099–10104
62. Kiefer, F., Tibbles, L.A., Anafi, M., Janssen, A., Zanke, B.W., Lassam, N., Pawson, T., Woodgett, J.R. and Iscove, N.N. (1996) EMBO J. **15**, 7013–7025
63. Hu, M.C.-T., Qiu, W.R., Wang, X., Meyer, C.F. and Tan T.-S. (1996) Genes Dev. **10**, 2251–2264
64 Su, Y.-C., Han, J., Xu, S., Cobb, M. and Skolnik, E.Y. (1997) EMBO J. **16**, 1279–1290

65. Pombo, C.M., Kehrl, J.H., Sánchez, I., Katz, P., Avruch, J., Zon, L.I., Woodgett, J.R., Force, T. and Kyriakis, J.M. (1995) Nature (London) **377**, 750–754
66. Ren, M., Zeng, J., De Lemos-Chiarandini, C., Rosenfeld, M., Adesnik, M. and Sabatini, D. (1996) Proc. Natl. Acad. Sci. U.S.A. **93**, 5151–5155
67. Chavrier, P., Vingron, M., Sander, C., Simons, K. and Zerial, M. (1990) Mol. Cell. Biol. **10**, 6578–6585
68. Vandenabeele, P., Declerck, W., Bayert, R. and Fiers, W. (1995) Trends Cell Biol. **5**, 392–399
69. Tartaglia, L.A. and Goeddel, D.V. (1992) Immunol. Today **13**, 151–153
70. Hsu, H., Xiong, J. and Goeddel, D.V. (1995) Cell **81**, 495–504
71. Hsu, H., Shu, H.-B, Pan, M.-G. and Goeddel, D.V. (1996) Cell **84**, 299–308
72. Rothe, M., Wong, S.C., Henzel, W.J. and Goeddel, D.V. (1994) Cell **78**, 681–692
73. Liu, Z.-G., Hsu, H., Goeddel, D.V. and Karin, M. (1996) Cell **87**, 565–576
74. Natoli, G., Costanzo, A., Ianni, A., Templeton, D.J., Woodgett, J.R., Balsano, C. and Levrero, M. (1997) Science **275**, 200–203
75. Stanger, B.Z., Leder, P., Lee, T.H., Kim, E. and Seed, B. (1995) Cell **81**, 513–523
76. Ting, A.T., Pimentel-Muiños, F.-X. and Seed, B. (1996) EMBO J. **15**, 6189–6195
77. Hsu, H., Huang, J., Shu, H.-B., Baichwal, V. and Goeddel, D.V. (1996) Immunity **4**, 387–396
78. Verheij, M., Bose, R., Lin, X.H., Yao, B., Jarvis, W.D., Grant, S., Birrer, M.J., Szabo, E., Zon, L.I., Kyriakis, J.M., et al. (1996) Nature (London) **380**, 75–79
79. Camard, C.J., Dbaibo, G.S., Liu, B., Obeid, L. and Hannun, Y.A. (1997) J. Biol. Chem. **272**, 16474–16481
80. Malinin, N.L., Boldin, M.P., Kovalenko, A.V. and Wallach, D. (1997) Nature (London) **385**, 540–544
81. Lavoie, J., L'Allemain, G., Brunet, A., Müller, R. and Pouysségur, J. (1996) J. Biol. Chem. **271**, 20608–20616
82. Xia, Z., Dickens, M., Raingeaud, J., Davis, R.J. and Greenberg, M.E. (1995) Science **270**, 1326–1331
83. Zanke, B.W., Boudreau, K., Rubie, E., Tibbles, L.A., Zon, L.I., Kyriakis, J., Liu, F.-F. and Woodgett, J.R. (1996) Curr. Biol. **6**, 606–613
84. Huang, S., Jiang, Y., Li, Z., Nishida, E., Mathias, P., Lin, S., Ulevitch, R.J., Nemerow, G.R. and Han, J. (1997) Immunity **6**, 739–749
85. Lenczowski, J.M., Dominguez, L., Eder, A.M., King, L.B., Zacharchuk, C.M. and Ashwell, J.D. (1997) Mol. Cell. Biol. **17**, 170–181
86. Goillot, E., Raingeaud, J., Ranger, A., Tepper, R.I., Davis, R.J., Harlow, E. and Sánchez, I. (1997) Proc. Natl. Acad. Sci. U.S.A. **94**, 3302–3307
87. Yang, X., Khosravi-Far, R., Chang, H. and Baltimore, D. (1997) Cell **89**, 1067–1076

Biochem. Soc. Symp. **64**, 49–62
Printed in Great Britain

4

Stress-activated MAP kinase (mitogen-activated protein kinase) pathways of budding and fission yeasts

Jonathan B.A. Millar

Division of Yeast Genetics, National Institute for Medical Research, The Ridgeway, Mill Hill, London NW7 1AA, U.K.

Abstract

All eukaryotic cells share the ability to sense and rapidly respond to environmental stress by initiating cyto-protective programmes of gene expression, protein translation and protein degradation. The molecular basis underlying these processes is, however, not well understood. Recently, attention has become focused on an evolutionarily conserved family of protein kinases called the stress-activated mitogen-activated protein kinases (SAPKs) that are activated when cells are challenged with a variety of environmental stresses or cytotoxic agents. Two members of the SAPK family, HOG1 and Sty1/Spc1, have been identified in the distantly related budding and fission yeasts, respectively. This has allowed researchers to genetically and biochemically dissect the structure of these pathways to begin to understand how they are activated and the role of the SAPKs in the cyto-protective response. In this chapter, I compare the structure of the SAPK pathways in the two yeasts and illustrate how this knowledge may benefit our understanding of stress sensing in mammalian cells.

Introduction

One of the most common responses of eukaryotic cells to growth-modulatory signals is the activation of one or more MAP (mitogen-

Nomenclature: standard fission yeast nomenclature has been used throughout the review. An example is as follows: the pathway is designated STY1, the gene *sty1*, cells bearing a wild-type gene *sty1*$^{+}$, mutant cells *sty1*$^{-}$ and the protein Sty1.

activated protein) kinase cascades. Signal transduction through MAP kinase cascades involves sequential phosphorylation and activation of three distinct kinases: the MAP kinase kinase kinase (or MAPKKK), the MAP kinase kinase (or MAPKK) and the MAP kinase itself. Although there appear to be multiple mechanisms by which plasma-membrane-associated receptors can induce activation of the MAPKKK, this initial activation leads to MAPKK activation by direct phosphorylation of two conserved serine or threonine residues in the catalytic domain of MAPKK. The MAPKK in turn activates the MAP kinase by dual phosphorylation on two closely spaced residues, a threonine and a tyrosine. Multiple MAP kinase cascades have been shown to operate in the same cell, each of which performs a different cellular function and is composed of distinct components (reviewed in [1–4]). Historically, much of the functional dissection of these signalling pathways has relied on genetic analysis in yeast. In particular, in this review I will refer to work on two evolutionarily divergent yeasts, namely the budding yeast *Saccharomyces cerevisiae* and the fission yeast *Schizosaccharomyces pombe*.

At least five distinct MAP kinase cascades have been identified in budding yeast that are variously required for mating, sporulation, control of cell polarity, pseudohyphal development and the cellular response to osmotic stress [1,4]. Although the molecular details of these pathways are not fully understood, analogous pathways for mating, cell wall construction and response to osmotic stress have been identified in fission yeast. Our present knowledge of the structure and function of the mammalian MAP kinase pathways is less well advanced, but the corresponding MAP kinases can be broadly classified into two groups: the extracellular-signal-regulated kinases (ERKs) and the stress-activated MAP kinases (SAPKs). The latter class is composed of at least seven distinct members and constitutes the largest group of MAP kinases in mammalian cells (see Chapter 3). The SAPKs have been implicated in numerous physiological and pathological conditions, including development, control of cell proliferation, cell death, inflammation and responses to ischaemic injury. As such, these enzymes are drawing considerable attention as potential targets for novel therapeutics. The mammalian SAPKs are activated by a variety of stress conditions (including osmotic stress, heat shock, oxidative stress, UV irradiation and the protein synthesis inhibitor anisomycin), as well as by inflammatory cytokines and certain vasoactive neuropeptides [5–10].

Two MAP kinases from budding yeast (Hog1) and fission yeast (Sty1) are members of the SAPK family. In this review, I describe a historical account of how the components of the HOG1 and STY1 pathways were identified, our present understanding of how they are activated and what functions they perform. I will compare and contrast the structures of these pathways with each other and with the SAPK pathways of metazoa. In this way I hope to convince the reader that the yeasts provide an excellent model system for understanding not only the structure and function of the eukaryotic SAPK pathways, but also more generally how cellular stress is sensed and dealt with by all eukaryotic cells.

The budding yeast HOG1 MAP kinase is activated by osmotic stress

Growth in all eukaryotic cells requires the uptake of water, driven by an osmotic gradient across the plasma membrane. When the external osmolarity increases, most cells are capable of osmoregulation by increasing their internal osmolarity. The budding yeasts respond to an increase in external osmolarity with increased synthesis of glycerol and decreased glycerol permeability, thereby increasing internal osmolarity. The founding member of the SAPK family, Hog1 (for **h**igh **o**smolarity **g**lycerol response), was initially identified in budding yeast from a genetic screen to isolate mutants that were unable to synthesize glycerol and proliferate under conditions of high osmolarity [11]. The Hog1 MAP kinase was found to be directly phosphorylated and activated by another gene product isolated from this screen, the Hog4 MAPKK [=MEK (MAP-kinase/ERK kinase)]. *hog4* is identical to *pbs2*, a gene that when overexpressed confers resistance to the antibiotic polymixin B [12]. Treatment of cells with osmotic stress leads to a rapid Hog4-dependent phosphorylation of Hog1. Hog1 is approx. 60% identical to the p38/CSBP1 subclass of stress-activated MAP kinases, and somewhat more distantly related (40–43% identity) to the SAPK/JNK (c-Jun N-terminal kinase) enzymes (Fig. 1). Despite this, both JNK1 and p38/CSBP (cytokine-suppressive anti-inflammatory drug-binding protein) 1 MAP kinases have been shown to rescue *hog1*$^{-}$ mutants when ectopically expressed in budding yeast [7,9]. Importantly, however, treatment with other forms of stress, including oxidative stress, heat shock and low pH, all failed to activate Hog1 [13]. Thus Hog1 appears to be specific for the cellular response to osmotic stress in budding yeast.

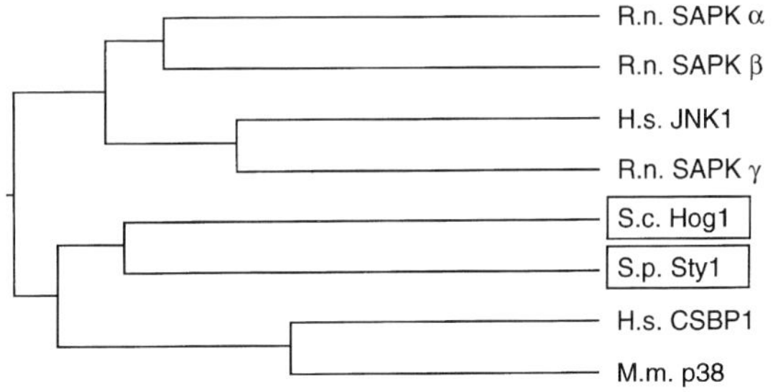

Fig. 1. Phylogenetic tree of the stress-activated MAP kinase family. Phylogenetic tree of amino acid sequence comparison of *Schizosaccharomyces pombe* (S.p.) Sty1 with *Rattus novegicus* (R.n.) SAPKαII, R.n. SAPKβ, R.n. SAPKγ [8], *Homo sapiens* (H.s.) JNK1 [5], *Saccharomyces cerevisiae* (S.c.) Hog1 [11], H.s. CSBP1 [9] and *Mus musculis* (M.m.) p38 [7]. The phylogenetic tree was generated using a MegAlign work package based on a Jotun Hein algorithm (DNASTAR).

Targets of the HOG1 kinase in the osmotic stress response

All eukaryotic cells exposed to mild stress develop tolerance not only against higher doses of the same stress (induced stress resistance), but also against stress caused by other agents (cross-protection), suggesting the existence of an integrating mechanism for sensing and responding to different forms of stress. In budding yeast, the promoters of multiple stress-inducible genes, including *ctt1*, *hsp12*, *gpd1* and *tps2*, contain a common regulatory element, known as the STRE (**st**ress **r**esponse **e**lement), which has the consensus core sequence AGGGG. The STRE is responsible for induction of these genes in response to multiple independent stresses. Mutations in either *hog1* or *hog4* abolish STRE-driven transcription in response to osmotic stress, but have no effect on induction of the same reporter in response to other environmental stresses, such as heat shock, oxidative stress or nitrogen starvation [13]. Msn2 and Msn4 are two structurally homologous and functionally redundant Cys_2His_2 zinc-finger transcription factors that have been found to bind to the STRE and mediate stress-induced transcription [14,15]. It seems likely that Msn2 and/or Msn4 are under the direct control of the Hog1 MAP kinase, but this has yet to be formally demonstrated. To date, no other potential targets of Hog1 have been identified.

Osmosensors that control HOG1 kinase activation

The screen for osmosensitive mutants in budding yeast did not reveal any upstream regulators of Hog4 or Hog1, suggesting that more than one gene product may control HOG4 function. A breakthrough in the identification of upstream regulators of the HOG pathway came from the isolation of two mutants, *ypd1*$^-$ and *ypd2*$^-$, that required overexpression of a tyrosine-specific phosphatase, *ptp2*, for viability [16]. Ptp2 has since been found to dephosphorylate and inactivate Hog1 directly (see below). Ypd2 was found to be identical to Sln1, an essential transmembrane histidine kinase structurally similar to a host of 'two-component systems' from bacteria [17–19]. A link to the HOG1 pathway was provided by the observation that mutants of either the *ssk4* (*hog1*) or *ssk3* (*hog4*) kinase bypass the lethality of an *sln1*$^-$ strain, suggesting that Sln1 negatively regulates Hog1. Loss of Sln1 can also be rescued by inactivation of two further genes, *ssk1* and *ssk2*, which code for a response regulator protein and a MAPKKK respectively, providing important additional links between Sln1 and Hog1 [20]. This exciting breakthrough gave the first indication that a two-component system may provide the means by which eukaryotic cells sense environmental stress. The biochemical details of the Sln1 two-component system have now been worked out in some detail [21]. Sln1 autophosphorylation initiates a four-step phospho-relay system in which a phosphate group is transferred from the Sln1 histidine kinase domain to an aspartate residue in the Sln1 response regulator domain and then to a histidine residue in the Ypd1 protein (see above). From Ypd1 the phosphate group is transferred to a conserved aspartate residue in the Ssk1 response regulator [21]. The unphosphorylated form of the Ssk1 response regulator binds the N-terminus of the Ssk2 MAPKKK, regulating its activity by an unknown mechanism [16,20] (Fig. 2). Ssk2 kinase activation leads to the sequential phosphorylation and activation of the Hog4 MAPKK and the Hog1 MAP kinase [11,16,20].

Despite our detailed knowledge of the two-component system in budding yeast, formal evidence that it actually transmits an osmotic stress signal is lacking. Importantly, inactivation of *ssk1* or simultaneous inactivation of *ssk2* and *ssk22* does not prevent activation of Hog1 in response to osmotic stress. To identify additional pathways that control Hog1, Maeda and colleagues [20] isolated mutants that were unable to respond to osmotic stress in the absence of *ssk2* and *ssk22*. In this manner a second transmembrane osmosensor, Sho1, was isolated [20]. In the absence of *ssk2*, *ssk22* and *sho1*, activation of the Hog1 kinase by osmotic stress is completely blocked. Sho1 encodes an Src homology-3 (SH3)-domain-containing protein which interacts directly with a proline-rich domain in the N-terminus of the Hog4 MAPKK (Fig. 2). Sho1 possesses no kinase activity by itself, but acts by recruiting a third MAPKKK, Ste11, to phosphorylate Hog4. The Hog4 MAPKK complexes with the Ste11 MAPKKK, Hog1 MAP kinase and Sho1 transmembrane protein, suggesting that Hog4 acts as a scaffold protein [22]. Curiously, Ste11 is

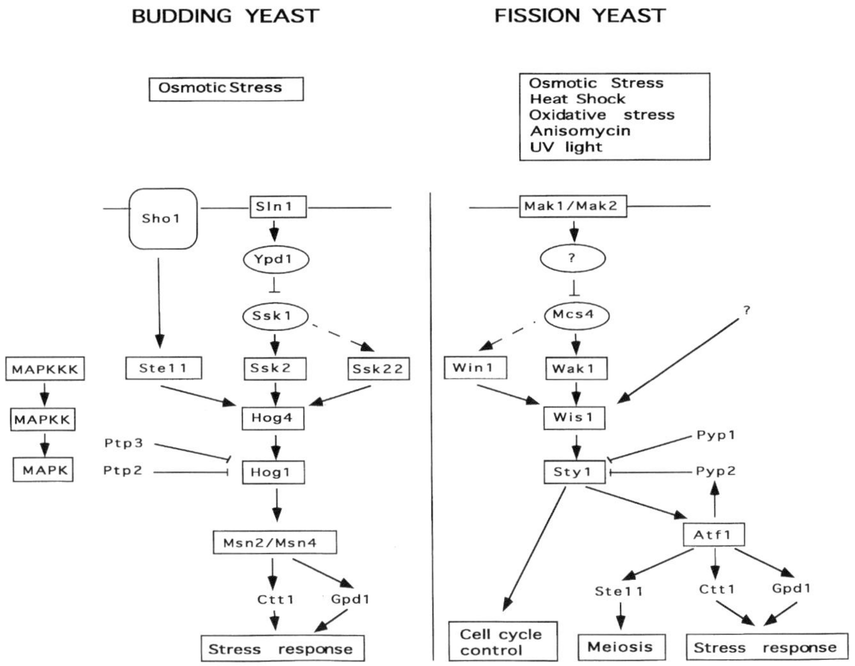

Fig. 2. Structures of the budding yeast HOG1 and fission yeast STY1 MAP kinase pathways.

the MAPKKK that phosphorylates the Ste7 MAPKK in response to mating pheromone. Despite this, the HOG1 and mating MAP kinase pathways appear to operate independently, since mating pheromone does not cause activation of Hog1 and osmotic stress does not cause activation of Fus3 or Kss1 [21]. The precise mechanism by which the Sho1→Ste11→Pbs2→Hog1 pathway is activated is at present unclear.

The fission yeast Sty1 MAP kinase is activated by multiple environmental stimuli

Since the fission yeast *Schizosaccharomyces pombe* is as evolutionarily divergent from budding yeast as it is from humans, comparative studies of cellular processes between the two yeasts have proved to be extremely informative. One exceptional example is the genetic analysis of cell cycle regulation, which has formed the basis of our understanding of eukaryotic cell proliferation and cancer aetiology in the past 10 years. It is now recognized that certain biological processes in mammalian cells appear to be operatively more similar to those found in fission yeast, whereas in other cases their function is more analogous to the situation in budding yeast. The stress-activated MAP kinase cascade appears to be an example of the former situation.

Three independent lines of investigation identified a homologue of Hog1 in fission yeast. First, mutations in two genes, *sty1* and *sty2*, were found to rescue simultaneous inactivation of two tyrosine-specific phosphatases, *pyp1* and *pyp2* [23]. Sty1 encodes a MAP kinase with high sequence similarity to the stress-activated MAP kinase family [23], while *sty2* was found to be allelic to *wis1*, which encodes a MAPKK that was previously identified as a mitotic inducer and a homologue of Hog4/Pbs2 ([24]; see below). The same genes were identified in a similar screen as suppressors of the lethal inactivation of two type 2C phosphatases, *ptc1* and *ptc3*, but in this case the MAP kinase was designated Spc1 [25,26]. Finally, Sty1 was also identified by PCR and called Phh1 [27], so that Sty1 is sometimes referred to as Spc1 or Phh1 in the literature, but for the purposes of this review I will use the Sty1 designation.

The Sty1 protein kinase is most closely related to Hog1 from budding yeast (86% identical); it is 57% identical to the p38 and CSBP1 MAP kinases from human and mouse cells [9] (Fig. 1), and somewhat more distantly related (between 40 and 43% identical) to a family of SAPKs from rat cells [8]. Notably, the residue separating the MAPKK phosphorylation sites in the catalytic domain appears to be conserved among members of the MAP kinase subfamilies. In human Erk1 and Erk2 and budding yeast Fus3 and Kss1 MAP kinases, this sequence is Thr-Glu-Tyr, whereas in members of the SAPK family and JNK1 it is Thr-Pro-Tyr, and in fission yeast Sty1, budding yeast Hog1, human CSBP1 and murine p38 it is Thr-Gly-Tyr. A phylogenetic tree compiled by sequence comparison with a number of stress-activated MAP kinases reveals that Sty1 belongs to the final subfamily of MAP kinases and has a Thr-Gly-Tyr motif (Fig. 1).

In contrast with the Hog1 kinase in budding yeast, Sty1 is activated by multiple environmental stresses, including osmotic stress, heat shock, peroxide and superoxide radicals, UV light, certain DNA-damaging agents and the protein synthesis inhibitor anisomycin [23,26,28–30]. In this manner there is a remarkable similarity in the agents that stimulate the fission yeast Sty1 kinase and the mammalian SAPK enzymes. That Sty1 is required for the cellular response to these stresses is illustrated by the fact that cells lacking Sty1 are unable to proliferate under conditions of high osmolarity or high temperature, or after exposure to UV irradiation [23,26,28–30]. In addition, cells lacking Sty1 are delayed in the timing of mitotic initiation and are unable to undergo sexual conjugation or differentiation (see below). Unlike budding yeast Hog1, therefore, the fission yeast Sty1 kinase is required for controlling multiple cellular events; to explain these various roles of Sty1, targets of the kinase have been sought.

Targets of the Sty1 kinase in the cellular response to stress

Several genes are induced in response to multiple stresses in fission yeast, including *pyp2* (encoding a MAP kinase phosphatase), *ctt1* (catalase) and *gpd1* (glyceraldehyde-3-phosphate dehydrogenase). Expression of these genes is completely abolished in the absence of either Sty1 MAP kinase or Wis1 MAPKK. In a search for transcription factors that may mediate this response, Atf1, a structural homologue of the human leucine-zipper-containing activating transcription factor 2 (ATF2), was found to be required for stress-induced transcription via the Sty1 MAP kinase pathway [31,32]. This is particularly striking, since the human SAPK1/JNK1 and p38/CSBP1 MAP kinases bind, phosphorylate and activate ATF2 [33]. A similar relationship was demonstrated between the Sty1 kinase and fission yeast Atf1; upon stimulation by stress, Atf1 binds to and becomes phosphorylated in its N-terminus by the Sty1 MAP kinase and, as a consequence, is transcriptionally activated [31,32] (Fig. 2). Mammalian ATF family members, including ATF1, bind directly to a consensus T(T/G)ACGTCA sequence variously known as the cAMP response element or ATF element. Several ATF elements have been identified in the *gpd1* and *ctt1* promoters, and direct binding of Atf1 to these promoters has been demonstrated [31]. Cells lacking *atf1* are partially sensitive to osmotic and heat stress, but not to UV irradiation, suggesting that additional Sty1 targets are required for the full stress response [29,31,32]. Nevertheless, these results indicate a remarkable similarity not only in the agents that activate the fission yeast and mammalian stress-activated MAP kinases but also in their transcriptional targets. Notably, there are no ATF family members in the whole budding yeast genome, indicating an important difference in the transcriptional targets of the budding yeast Hog1 and fission yeast Sty1 MAP kinases.

The Wis1→Sty1→Atf1 pathway also controls sexual differentiation

The fission yeast Atf1 transcription factor was originally described as a gene product required for entry into G1 phase and sexual conjugation under poor nutrient conditions [34,35]. Cells lacking the Sty1 kinase or the Wis1 MAPKK are also unable to arrest in G1 or to mate, suggesting that phosphorylation of Atf1 by Sty1 is also required for its meiotic role [31,32,34]. One of the major targets in the sexual differentiation pathway is the high-mobility group-box-containing transcription factor Ste11. Ste11 plays a central role, being required for expression of the mating MAP kinase pathway and the subsequent response to mating pheromone. Induction of *ste11* mRNA in response to nitrogen starvation is completely abolished in cells lacking *wis1*, *sty1* or *atf1* and, as a consequence, cells are unable to mate. These data indicate a crucial role for the stress-activated MAP kinase pathway in cellular differentiation in this organism.

In contrast with a recent report [32], mounting evidence suggests that the Sty1 MAP kinase is not activated during nitrogen starvation or by mating pheromone (M.G. Wilkinson, M. Samuels, N.C. Jones and J.B.A. Millar, unpublished work). Certain genes (e.g. *ctt1* and *pyp2*) are induced by stress in an Atf1-dependent manner, but not by nitrogen starvation, whereas others (e.g. *fbp1* and *ste11*) are induced by nitrogen starvation in an Atf1-dependent manner, but not by stress ([31,32,36]; M.G. Wilkinson and J.B.A. Millar, unpublished work). How is it that a single transcription factor can be involved in two different cellular responses when its function is dependent on phosphorylation by the same MAP kinase in both cases? A clue to this probably lies in the fact that there are at least four ATF family members in fission yeast, each of which can form multiple homodimeric and heterodimeric complexes; distinct complexes appear to have different promoter target specificities ([32,34,35,37]; M.W. Toone, M. Samuels, J.B.A. Millar and N.C. Jones, unpublished work). Intriguingly, one binding partner for Atf1, Pcr1, is required for sexual differentiation, but plays no role in the cellular stress response [37]. It is conceivable that differently phosphorylated forms of Atf1 may have altered affinities for other ATF family members and that switches in binding partner can be triggered by altering the nutrient conditions, thus changing the ultimate promoter target specificity. This hypothesis is supported by the observation that Atf1 undergoes a rapid dephosphorylation in response to nutrient limitation, and that, under these conditions, Atf1 preferentially binds to Pcr1 [35]. It should be noted that Atf1 is still phosphorylated in nitrogen-starved cells by the Sty1 MAP kinase, but to a much lesser extent than in exponentially growing cells [35]. A switch in ATF binding partner is most likely catalysed by the drop in intracellular cAMP which accompanies nitrogen starvation in fission yeast, rather than by a change in Sty1 activity.

Sty1 kinase controls the timing of mitotic initation

The fission yeast cell cycle is controlled at two major points: in G1 at entry into S phase (DNA replication) and in G2 at the initiation of mitosis. Genetic and physiological studies have revealed that the timing of both transitions requires attainment of a critical cell size [38]. Under conditions of normal growth, the mass sensor governing the initiation of DNA synthesis is cryptic, such that cell size at division is rate-limiting for cell cycle progression. Since *S. pombe* cells grow by length extension, this is expressed as the attainment of a critical cell length. It is now recognized that mitotic initiation is triggered by activation of the catalytic kinase subunit of the Cdc13 (cyclin B)/Cdc2 kinase. Genes that when mutated alter cell size at division are, by inference, required for the correct timing of Cdc2 activation. Two such genes are the *wee1* and *cdc25* mitotic regulators, which code for a tyrosine kinase and phosphatase respectively that control directly the activity of the Cdc13–Cdc2 complex [39,40]. Importantly, cells lacking *wis1* or *sty1* are also delayed in the timing of the G2/M transition and display strong genetic interactions with the *cdc25* phosphatase. These observations point to a crucial link between the cellular response to environmental stress and control of the cell cycle [23–26]. The obvious question to ask is: how does the Sty1 kinase control cell cycle progression? Genetic evidence suggests that the control over the G2/M transition by the Sty1 MAP kinase pathway is independent of Wee1 and Cdc25, and may act to regulate either the formation and/or the localization of the Cdc13–Cdc2 complex. Cells depleted of the Atf1 transcription factor have no defect in cell size at division, suggesting that an alternative substrate(s) of Sty1 mediates this control [31,32].

Upstream regulators of the Sty1 MAP kinase

Since the fission yeast Sty1 MAP kinase and the mammalian stress-activated SAPK/JNK and p38/CSBP1 MAP kinases are activated by a range of environmental insults, it is reasonable to suppose that the upstream regulators of the mammalian and fission yeast pathways may be structurally similar. We have recently identified an upstream regulator of Sty1 as the product of another cell cycle regulator, *mcs4* [30,34]. Mcs4 is structurally and functionally homologous to the budding yeast Ssk1 response regulator, indicating that the fission yeast Sty1 kinase, like the budding yeast Hog1 kinase, is controlled by a conserved two-component system [30]. Mcs4 acts upstream of Wak1, a homologue of the Ssk2 and Ssk22 MAPKKKs, which transmits the stress signal to the Wis1 MAPKK [30,41]. A second MAPKKK with homology to Wak1 and the Ssk2 and Ssk22 kinases has been identified as the product of another cell cycle regulator, *win1*, although it is not known whether this kinase is also regulated by Mcs4 ([42]; I. Samejima and P. Fantes, personal communication). Loss of *mcs4* attentuates the induction of gene transcription in response to multiple stresses, including osmotic stress, oxidative stress, heat shock and anisomycin, suggesting that the two-component system may sense multiple stresses [30]. We have identified two functionally overlapping transmembrane histidine kinases, Mak1 and Mak2, with structural similarity to budding yeast Sln1 that control Mcs4 function. However, further analysis has suggested that the fission yeast two-component system is not the primary stress sensor for the Sty1 MAP kinase cascade, but may be required for sensing perturbation of the cell wall architecture (V. Buck, K. Makino and J.B.A. Millar, unpublished work).

If this hypothesis is true, then there must be alternative stress sensors in fission yeast. Consistent with this idea, we have recently uncovered a novel pathway controlling Wis1 MAPKK function that is independent of a MAPKKK. Specifically, Wis1 is rapidly phosphorylated *in vivo* in response to environmental stress in the absence of the Wak1/Wik1 or Win1 MAPKKKs, and on residues distinct from Ser-469, Ser-471 and Thr-473. Deletion analysis suggests that this phosphorylation occurs in the N-terminal non-catalytic domain of Wis1 and that this domain is important for Wis1 activity. Moreover, we have shown that phosphorylation of Wis1 in response to stress is not due to autophosphorylation or feedback inhibition by Sty1 MAP kinase (J.-C. Shieh and J.B.A. Millar, unpublished work). The functional consequences of this phosphorylation for the activity and/or localization of the Wis1 MAPKK *in vivo* have yet to be determined. On the face of it, this may seem analogous to the Sho1/Ste11 arm of the HOG1 pathway, but differs significantly in that involvement of an Ssk2 or Ste11 homologue has been ruled out. Indeed, these results indicate that stress-induced activation of the Sty1 MAP kinase requires phosphorylation of the non-catalytic domain of the Wis1 MAPKK by an as-yet-unidentified protein kinase (Fig. 2).

Negative regulators of the HOG1 and STY1 cascades

The amplitude and duration of signal transmission through the MAP kinase cascade is tightly controlled by a network of opposing kinases and phosphatases. A number of phosphatases have been identified that negatively regulate the HOG1 and STY1 pathways. These can be grouped into two classes: the protein-tyrosine phosphatases and the type 2C serine/threonine phosphatases. In fission yeast the Pyp1 and Pyp2 protein-tyrosine phosphatases act as MAP kinase phosphatases by specifically dephosphorylating Tyr-173 of the Sty1 MAP kinase [23,26]. All other MAP kinase phosphatases described to date dephosphorylate both threonine and tyrosine residues, and thus have dual specificity [43–46]. This suggests that Pyp1 and Pyp2 are the founding members of a new class of single-specificity MAP kinase phosphatases, and recent evidence suggests that *S. cerevisiae* Ptp2 and Ptp3 play a similar role for Hog1 [47]. Pyp1 and Pyp2 are differentially regulated, in that the *pyp2* but not *pyp1* mRNA is induced by stress in a manner requiring the Wis1→Sty1→Atf1 pathway [31]. Thus Pyp2 participates in a negative-feeback loop to regulate Sty1 activity, while the function of the Pyp1 MAP kinase phosphatase is at present uncertain. The identity of the phosphatase(s) that dephosphorylates Thr-171 also remains unknown. Overexpression of any of three type 2C serine/threonine phosphatases (Ptc1, Ptc2, Ptc3) in fission yeast bypasses lethal hyperactivation of the Sty1 pathway [25]. The target of these phosphatases has yet to be determined, but appears to act downstream of the MAP kinase [47a]. It remains to be determined whether protein-tyrosine phosphatases or type 2C serine/threonine phosphatases regulate the mammalian SAPK pathways.

Conclusions and unanswered questions

As we have seen in this chapter, there are some striking similarities in the architecture of the stress-activated budding yeast HOG1 and fission yeast STY1 MAP kinase pathways. Both yeast pathways are controlled by a two-component system that acts by modulating the activity of one of two MAPKKKs. Despite this, budding yeast Hog1 is activated only by osmotic stress, whereas fission yeast Sty1 is activated by multiple stresses. Evidence from the entire genome sequence of *S. cerevisiae* indicates that Sln1 is the only histidine kinase in that organism. Thus it is conceivable that fission yeast contains multiple histidine kinases, each of which responds to a different stress. Although we have identified at least two histidine kinases in fission yeast, our evidence suggests that the two-component system is not the primary stress sensor for the STY1 pathway. Instead, initial evidence suggests that the fission yeast two-component system is specifically sensing perturbation of cell wall integrity or turgor pressure. It is notable that the eukaryotic two-component system has been identified only in eukaryotic cells that possess a cell wall, namely yeasts, plants and slime moulds. Indeed since no two-component system has been identified in mammalian cells, despite extensive searching, it is tempting to speculate that this is because mammalian cells do not possess a cell wall.

So how are the mammalian SAPK pathways activated? Despite the identification of at least six MAPKKKs that regulate the SAPK/JNK pathway, including MEK kinase 1 (MEKK1), MUK, mixed-lineage kinases (MLK1, MLK2 and MLK3), transforming growth factor-β-activated kinase (TAK) and apoptosis-stimulating kinase 1 (ASK1), none has been shown to be activated by environmental stress [48–53], so that the mechanism by which the mammalian stress-activated MAP kinase pathways are activated remains a mystery. A clue to how this apparent paradox may be solved has again come from studies in yeast. Activation of Hog4 MAPKK is stimulated by the recruitment of the Ste11 MAPKKK by the SH3-domain-containing transmembrane protein Sho1, in the absence of a two-component system. In this case the N-terminal non-catalytic domain of Hog4 is crucial for complex-formation and signal transmission. Although the details of this activation have yet to be worked out, it is interesting that the N-terminus of the fission yeast Wis1 MAPKK is subject to novel stress-activated phosphorylation that is independent of a MAPKKK or the two-component system (J.-C. Shieh and J.B.A. Millar, unpublished work). It may be that this is an evolutionarily conserved mechanism by which the mammalian SAPK pathways are activated. Recruitment of the MAPKK into a complex containing a MAPKKK would alleviate the need for the MAPKKK to be catalytically stimulated in order to induce activation of the MAP kinase. Further experimentation will be needed to test whether this model is correct.

Despite the identification of a number of components of the stress-activated MAP kinase cascade in both budding yeast and fission yeast, we are still far from our goal of fully understanding how these pathways operate. Nevertheless, I hope I have convinced the reader that the yeasts are excellent model systems in which not only to study the eukaryotic SAPK pathways but more generally to determine how stress is sensed by eukaryotic cells and how they transduce and respond to this information.

I thank members of the Division of Yeast Genetics for advice and critical reading of the manuscript. J.B.A.M. is supported by the MRC (U.K.)

References

1. Ammerer, G. (1994) Curr. Opin. Genet. Dev. **4**, 90–95
2. Davis, R.J. (1994) Trends Biochem. Sci. **19**, 470–473
3. Marshall, C.J. (1994) Curr. Opin. Genet. Dev. **4**, 82–89
4. Herskowitz, I. (1995) Cell **80**, 187–197
5. Dérijard, B., Hibi, M., Wu, I.-H., Barrett, T., Su, B., Deng, T., Karin, M. and Davis, R.J. (1994) Cell **76**, 1025–1037
6. Galcheva-Gargova, Z., Dérijard, B., Wu, I.-H. and Davis, R.J. (1994) Science **265**, 806–808
7. Han, J., Lee, J.-D., Bibbs, L. and Ulevitch, R.J. (1994) Science **265**, 808–811
8. Kyriakis, J.M., Banerjee, P., Nikolakaki, E., Dai, T., Rubie, E.A., Ahmad, M.F., Avruch, J. and Woodgett, J.R. (1994) Nature (London) **369**, 156–160
9. Lee, J.C., Laydon, J.T., McDonnell, P.C., Gallagher, T.F., Kumar, S., Green, D., McNulty, D., Blumenthal, M.J., Heys, J.R., Landvatter, S.W., et al. (1994) Nature (London) **372**, 739–746
10. Rouse, J., Cohen, P., Trigon, S., Morange, M., Alonso-Llamazares, A., Zamanillo, D., Hunt, T. and Nebreda, A. (1994) Cell **78**, 1027–1037
11. Brewster, J.L., de Valoir, T., Dwyer, N.D., Winter, E. and Gustin M.C. (1993) Science **259**, 1760–1763
12. Boguslawski, G. and Polazzi, T. (1987) Proc. Natl. Acad. Sci. U.S.A. **84**, 5848–5852
13. Schüller, C., Brewster, J.L. Alexander, M.R., Gustin, M.C. and Ruis, H. (1994) EMBO J. **13**, 4382–4389

14. Martinez-Pastor, M.T., Marchler, G., Schüller, C., Marchler-Bauer, A., Ruis, H. and Estruch, F. (1996) EMBO J. **15**, 2227–2235
15. Schmitt, A.P. and McEntee, K. (1996) Proc. Natl. Acad. Sci. U.S.A. **93**, 5777–5782
16. Maeda, T., Wurgler-Murphy, S.M. and Saito, H. (1994) Nature (London) **369**, 242–245
17. Stock, J.B., Ninfa, A.J. and Stock, A.M. (1989) Microbiol. Rev. **53**, 450–490
18. Parkinson, J.S. (1993) Cell **73**, 857–871
19. Ota, I.M. and Varsharvsky, A. (1993) Science **262**, 566–569
20. Maeda, T., Takekawa, M. and Saito, H. (1995) Science **269**, 554–558
21. Posas, F., Wurgler-Murphy, S.M., Maeda, T., Witten, E.A., Thai, T.C. and Saito, H. (1996) Cell **86**, 865–875
22. Posas, F. and Saito, H. (1997) Science **276**, 1702–1705
23. Millar, J.B.A., Buck, V. and Wilkinson, M.G. (1995) Genes Dev. **9**, 2117–2130
24. Warbrick, E. and Fantes, P. (1991) EMBO J. **10**, 4291–4299
25. Shiozaki, K. and Russell, P. (1995) EMBO J. **14**, 492–502
26. Shiozaki, K. and Russell, P. (1995) Nature (London) **378**, 739–743
27. Kato, T., Okazaki, K., Murakami, H., Stettler, S., Fantes, P. and Okayama, H. (1996) FEBS Lett. **378**, 207–212
28. Degols, G., Shiozaki, K. and Russell, P. (1996) Mol. Cell. Biol. **16**, 2870–2877
29. Degols, G. and Russell, P. (1997) Mol. Cell. Biol. **17**, 3356–3363
30. Shieh, J.-C., Wilkinson, M.G., Buck, V., Morgan, B., Makino, K. and Millar, J.B.A. (1997) Genes Dev. **11**, 1008–1022
31. Wilkinson, M.G., Samuels, M., Takeda, T., Toda, T., Toone, M.W., Shieh, J.-C., Millar, J.B.A. and Jones, N.C. (1996) Genes Dev. **10**, 2289–2301
32. Shiozaki, K. and Russell, P. (1996) Genes Dev. **10**, 2276–2288
33. Gupta, S., Campbell, D., Derijard, B. and Davies, R.J. (1995) Science **267**, 389–393
34. Takeda, T., Toda, T., Kominami, K., Kohnosu, A., Yanagida, M. and Jones, N. (1995) EMBO J. **14**, 6193–6208
35. Kanoh, J., Watanabe, Y., Ohsugi, M., Lino, Y. and Yamamoto, M. (1996) Genes Cells **1**, 391–408
36. Stettler, S., Warbrick, E., Prochnik, S., Mackie, S. and Fantes, P. (1996) J. Cell Sci. **109**, 1927–1935
37. Watanabe, Y. and Yamamoto, M. (1996) Mol. Cell. Biol. **16**, 704–711
38. Nurse, P. (1975) Nature (London) **256**, 547–551
39. Russell, P. and Nurse, P. (1987) Cell **49**, 559–567
40. Millar, J.B.A. and Russell, P. (1992) Cell **68**, 407–410
41. Shiozaki, K., Shiozaki, M. and Russell, P. (1997) Mol. Biol. Cell. **8**, 409–419
42. Ogden, J.E. and Fantes, P. (1986) Curr. Genet. **10**, 509–514
43. Keyse, S.M. and Emslie, E.A. (1992) Nature (London) **359**, 644–647
44. Sun, H., Charles, C.H., Lau, L.F. and Tonks, N. (1993) Cell **75**, 487–493
45. Doi, K., Gartner, A., Ammerer, G., Errede, B., Shinkawa, H., Sugimoto, K. and Matsumoto, K. (1994) EMBO J. **13**, 61–70
46. Ward, Y., Gupta, S., Jensen, P., Wortmann, M., Davis, R.J. and Kelly, K. (1994) Nature (London) **367**, 651–654
47. Wurgler-Murphy, S.M., Maeda, T., Witten, E.A. and Saito, H. (1997) Mol. Cell. Biol. **17**, 1289–1297
47a. Gaits, F., Shiozaki, K. and Russell, P. (1997) J. Biol. Chem. **272**, 17873–17879
48. Yan, M., Dai, T., Deak, J.C., Kyriakis, J.M., Zon, L.I., Woodgett, J.R. and Templeton, D.J. (1994) Nature (London) **372**, 798–800
49. Yamaguchi, K., Shirakabe, K., Shibuya, H., Irie, K., Oishi, I., Ueno, N., Taniguchi, T., Nishida, E. and Matsumoto, K. (1995) Science **270**, 2008–2011

50. Hirai, S.-i., Izawa, M., Osada, S.-i., Spyrou, G. and Ohno, S. (1996) Oncogene **12**, 641–650
51. Rana, A., Gallo, K., Godowski, P., Hirai, S., Ohno, S., Zon, L., Kyriakis, J.M. and Avruch, J. (1996) J. Biol. Chem. **271**, 19025–19028
52. Salmeron, A., Ahmad, T.B., Carlile, G.W., Pappin, D., Narsimhan, R.P. and Ley, S.C. (1996) EMBO J. **15**, 817–826
53. Ichijo, H., Nishida, E., Irie, K., ten-Dijke, P., Saitoh, M., Moriguchi, T., Takagi, M., Matsumoto, K., Miyazono, K. and Gotoh, Y. (1997) Science **275**, 90–94

Biochem. Soc. Symp. **64**, 63–77
Printed in Great Britain

5

Protein kinase cascades in intracellular signalling by interleukin-1 and tumour necrosis factor

Jeremy Saklatvala*, Jon Dean and Andrew Finch

Division of Cell Signalling, Kennedy Institute of Rheumatology, 1 Aspenlea Road, Hammersmith, London W6 8LH, U.K.

Abstract

Interleukin 1 (IL-1) and tumour necrosis factor (TNF) are major mediators of inflammation, with similar actions. Their receptor mechanisms and downstream pathways are reviewed. They activate several protein kinases in fibroblasts, including the three types of mitogen-activated protein kinase (MAPK), the kinase of the inhibitor of nuclear factor-κB (IκBK), and the TNF-/IL-1-activated β-casein kinase. Cultured cells show a broader spectrum of kinase activation by IL-1 than tissues *in vivo*, suggesting that the receptors connect to more pathways in proliferating cells than in resting differentiated cells. The c-Jun N-terminal kinase (JNK) is strongly activated by IL-1 in tissues. In rabbit liver this is mediated by MAPK kinase 7; the upstream kinase is unidentified. Little is known of downstream MAPK targets in inflammation. Inhibitor experiments suggest that p38MAPK mediates induction of cyclo-oxygenase-2 and metalloproteinases by IL-1, and of TNF, IL-1 and cyclo-oxygenase-2 by endotoxin (in monocytes). p38MAPK is needed for induction of the mRNAs (except IL-1 mRNA).

Introduction: biological action of the cytokines

Interleukin-1α (IL-1α), IL-1β and tumour necrosis factor α (TNFα) are primary inflammatory cytokines that mediate many of the local and systemic features of inflammation [1,2]. They are produced mainly by monocytes and macrophages in response to a range of stimuli, including microbial products, immune complexes, activated T-cells and the

*To whom correspondence should be addressed.

combined action of other cytokines, such as interferon γ, IL-2 and granulocyte/macrophage colony-stimulating factor. TNFβ, or lymphotoxin α, is made by activated T-lymphocytes. The IL-1s and TNFs have a highly similar, broad range of physiological effects and provide crucial signals by which activated mononuclear phagocytes control functions of other cells. The most notable biological difference between them is that the TNFs have cytotoxic properties, whereas the IL-1s do not.

When they are produced at sites of tissue damage or microbial invasion, these two cytokines (IL-1 and TNF) act as distress signals to recruit leucocytes and to activate defence mechanisms in distant organs. They cause local accumulation of leucocytes by stimulating production of chemotactic factors (e.g. IL-8, monocyte chemotactic protein, platelet-activating factor etc.) and by inducing the expression of adhesion molecules on vascular endothelial cells. The up-regulation of molecules such as E-selectin and intercellular adhesion molecule-1 on the luminal surface of the endothelial cells causes leucocytes to adhere to this surface. Subsequently the leucocytes move out into the extravascular space in response to the chemotactic stimuli.

IL-1 and TNF stimulate connective tissue cells and vascular endothelial cells to make prostaglandins, which cause other local effects such as increased blood flow and capillary permeability, and enhanced perception of pain. They also stimulate connective tissue cells to resorb the extracellular matrix, and their continued production in chronic inflammatory diseases results in tissue destruction. An example of this is the destruction of cartilage, bone and tendons in rheumatoid arthritis.

IL-1 and TNF act on distant target organs partly directly and partly via the generation of secondary cytokines such as IL-6. They stimulate the production of acute-phase plasma proteins in the liver, and they are very potent pyrogens. They cause fever by stimulating prostaglandin production in the pre-optic area of the hypothalamus. This also stimulates the production of corticotrophin-releasing factor, which in turn stimulates the pituitary–adrenal axis.

Underlying the complex physiology of IL-1 and TNF is their ability to induce the expression of many genes. The promoters of these genes are controlled in a complex manner by transcription factors such as nuclear factor-κB (NF-κB), activator protein-1 (AP-1), activating transcription factor 2 (ATF2) and nuclear factor interleukin-6. The activity of these factors is modulated by phosphorylation. IL-1 and TNFα have been shown to activate essentially the same sets of protein kinases. These are arranged in phosphorylation cascades, which presumably serve to diversify and amplify the signals.

Protein kinases activated by IL-1 and TNF: differences between cultured cells and tissues *in vivo*

The kinases activated by IL-1 and TNF are summarized in Fig. 1. They include the three types of mitogen-activated protein kinase (MAPK)

[3–8], a kinase cascade leading to phosphorylation and degradation of IκB, the inhibitor of NF-κB [9,10], and a little-understood enzyme that is highly specific to IL-1 and TNF which phosphorylates a β-casein peptide and has been called TNF- and IL-1-activated protein kinase (TIPK)

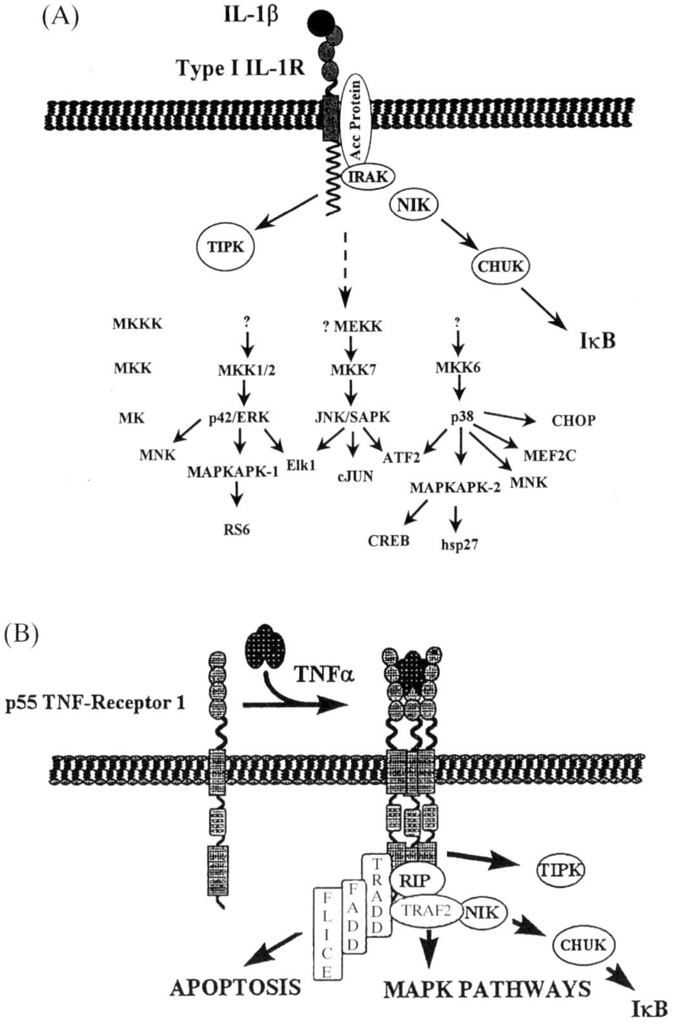

Fig. 1. IL-1 type 1 (A) and TNF p55 (B) receptor-associated proteins and downstream protein kinase pathways. The MAPK pathways are not detailed in (B). The abbreviations of the receptor-associated proteins are explained in the text. MK is the same as MAPK. MKK is MAPK kinase. MKKs are also called MEKs (MAP-kinase/ERK kinases). MKKKs are therefore also called MEKKs. MAPKAPK-1 is MAP-kinase-activated protein kinase 1, also known as RSK [ribosomal S6 protein (RS6) kinase]. Other abbreviations: IL-1R, IL-1 receptor; NIK, NF-κB-activating kinase.

[3,11,12]. Apart from TIPK, these enzymes are not specific for the cytokines. The original p42/p44 MAPK is activated by very many stimuli, including mitogens, and the c-Jun N-terminal kinase (JNK)/stress-activated protein kinase (SAPK) and p38 enzymes are activated by many cellular stresses and by microbial products such as lipopolysaccharide (LPS), in addition to IL-1 and TNF.

While activation of the MAPKs by the cytokines in various cultured cells is well established, the links upstream to the cytokine receptors are still uncertain. It is also becoming clear that the range of kinases activated varies considerably with the type of cell being studied. When cultured fibroblasts are stimulated by either cytokine, there is activation of all three MAPK pathways. Transformed cell lines such as HeLa and KB show a similar pattern of activation to fibroblasts, although p42/p44 MAPK activation is less pronounced [8]. Tissues *in vivo* react differently. We have injected IL-1 into the systemic circulation of rabbits and then made homogenates of tissues 10–15 min post-injection. These were assayed for the three types of MAPK by immunoprecipitating the enzymes and measuring their activity with appropriate substrates. So far we have only observed activation of JNK/SAPK, and not of p42/p44 MAPK or p38. Figure 2(A) shows the results for the lung. This tissue comprises a large capillary bed, and the main cell type stimulated by the cytokine will be the vascular endothelial cell. Some signals may also arise from the epithelial cells which form the alveolar lining. Thus resting differentiated vascular endothelial cells appear to show only JNK/SAPK activation. Figure 2(B)

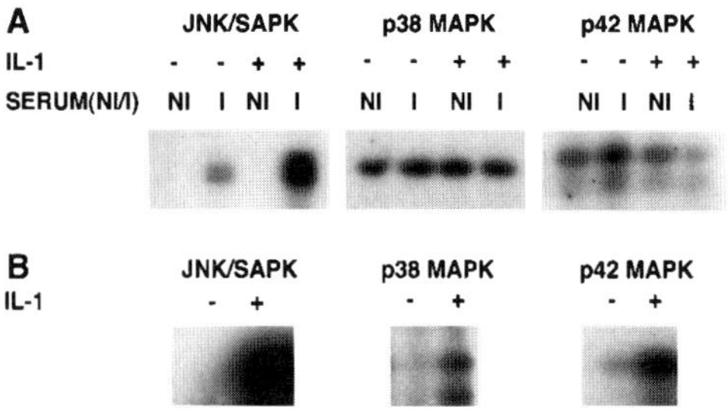

Fig. 2. Activation of MAPK pathways by IL-1. (A) IL-1 activates JNK/SAPK, but not p42 or p38 MAPKs, in rabbit lung. Rabbits were injected with vehicle or IL-1, and organs were removed (10 min post-IL-1) after lethal injection. Immunoprecipitates were made from lung lysates with non-immune (NI) or immune (I) serum against the relevant kinase. Rabbit antisera were to (i) a C-terminal peptide of SAPKα, (ii) a C-terminal peptide of p38 MAPK or (iii) a C-terminal peptide of p42 MAPK [13]. Immunoprecipitates were incubated with assay buffer containing [γ-^{32}P]ATP and the relevant substrate: GST–Jun-(1–135) for JNK/SAPK, [His]MAPKAPK-2 and Hsp27 for p38 MAPK, and myelin basic protein for p42 MAPK. Phosphorylated substrates were visualized by a Phosphorimager after SDS/PAGE. (B) IL-1 activates all three pathways in human umbilical vein endothelial cells.

shows an experiment carried out on cultured vascular endothelial cells (human rather than lapine). In these, IL-1 activates all three pathways, as judged by increased phosphorylation by cell lysates of an epidermal growth factor receptor peptide containing Thr-669 (a JNK substrate), myelin basic protein (the substrate for p42/p44 MAPK) and heat-shock protein of 27 kDa (Hsp27; a substrate downstream of p38) (Fig. 2B). We have seen a similar pattern of activation of all three MAPKs by IL-1 in fibroblasts and chondrocytes. We have also investigated tissues other than the lung. Both liver [13] and spleen show strong activation of JNK/SAPK and little or no change in p38 or p42/p44 MAPK.

Further work is needed to establish which forms of JNK/SAPK are being activated in cells and tissues. The immunoprecipitating antibody used in the experiment shown in Fig. 2 was made to the C-terminal sequence of rat SAPKα (equivalent to human JNK2) and cross-reacts with the C-terminus of SAPKβ (JNK3). We formerly purified the two major forms of SAPK from IL-1-treated rabbit liver [7]. Together these accounted for at least 90% of the activity in the tissue. They had molecular masses of 55 and 50 kDa, and we sequenced a number of peptides from each form. All the peptides corresponded precisely to the SAPKα sequence. There were no β or γ peptides. We have used other antibodies made to the γ and β sequences (which do not recognize α), and these precipitated activated JNK/SAPK from cultured fibroblasts and HeLa cells, but not from liver. These results indicate that there may be tissue- or cell-specific activation or expression of the different forms of JNK/SAPK.

We have not yet investigated whether there is differential expression or activation of different forms of p38 in different cells. Our immunoprecipitation assay measures the original p38 [5]; the antiserum does not recognize p38β/p38-2 [14,15] or the less closely related enzymes p38γ (also called SAPK3 and ERK6, where ERK stands for extracellular-signal-regulated kinase) [16] and p38δ (SAPK4) [17,18].

In cultured cells, both transformed (HeLa, KB) and untransformed (fibroblasts, endothelial cells, chondrocytes), IL-1 and TNF also activate the dual-specificity TIPK whose activity we originally discovered with a β-casein substrate [11]. It phosphorylates Ser-124 in β-casein and favours hydrophobic residues at positions +1, +9 and +11. Its physiological substrates are unknown, but it also phosphorylated the β-casein peptide when Ser-124 was replaced by tyrosine [12]. We have not been able to identify this kinase in tissues *ex vivo*, only in cultured cells. Thus IL-1 activates fewer protein kinases in tissues (mainly JNK/SAPK) than in cells in culture (all three MAPKs and TIPK). Why should this be so? One can speculate that cultured cells, especially primary cultures of fibroblasts, vascular endothelial cells and chondrocytes, are akin to those involved in tissue repair or wound healing. Such cells carry out functions distinct from those of their resting differentiated counterparts, e.g. cell division, movement, rapid remodelling of the extracellular matrix, organization of different cellular elements into a granulation tissue, and so on. These processes would need to be under cytokine control, so such cells would need more signalling pathways through which the cytokines could operate than would their resting precursors.

MAPK kinases (MKKs) activated by IL-1 and TNF

The results described above suggest the possibility that the connection of the cytokine receptors with downstream kinases may vary between cells or with the state of differentiation of the cell. Our understanding of these connections is incomplete, as shown in Fig. 1. A number of MAPK activators, or MKKs, have been cloned and show a high degree of specificity. MKK1 and MKK2 are closely related enzymes that activate the p42/p44 MAPKs. In cells such as fibroblasts, in which these are activated by IL-1 and TNF, this occurs via MKK1 (and probably MKK2) [19]. The nature of the kinase upstream of MKK1/2 is not yet clear. Growth factors that utilize tyrosine kinase receptors activate c-Raf, and this phosphorylates and activates MKK1/2. However, we have been unable to show activation of c-Raf in either IL-1- or TNF-stimulated fibroblasts [20], although there have been reports of TNF activating c-Raf [21].

The activator of p38 MAPK used by IL-1 in cultured cells appears to be the same as that used by stressful stimuli, and has been identified on the basis of immunoreactivity as MKK6 [22,23]. MKK6 was originally identified as a cDNA similar to MKK3 [24], the other p38 activator so far identified. MKK6 has also been purified from skeletal muscle, and is likely to be a physiological activator in this tissue [23].

An IL-1-induced activator of JNK/SAPK closely related to MKK7

When IL-1 was injected into rabbits, it caused marked activation of JNK/SAPK in several tissues. We tried to identify an activator of JNK/SAPK in liver [13]. To assay an activator, the rat SAPKβ form of JNK/SAPK was expressed as a glutathione S-transferase (GST) fusion protein for use as a substrate. This was incubated with chromatography fractions or tissue extract in the presence of ATP, then absorbed on to GSH–agarose beads. These were washed well, and the immobilized enzyme was assayed for its ability to phosphorylate a fragment of c-Jun. This comprises the first 135 amino acids and contains Ser-63 and Ser-73, which are the sites phosphorylated by JNK/SAPK when it activates the transcription factor.

Figure 3 shows that, with this assay, it was possible to detect an IL-1-regulated activator in fractions from cation-exchange chromatography. Unfortunately this activator did not bind to anion-exchangers, and it eluted rather broadly from cation-exchangers developed with salt gradient elution. Although we were able to purify it several hundred fold by sequential chromatography, we were not able to identify the kinase as

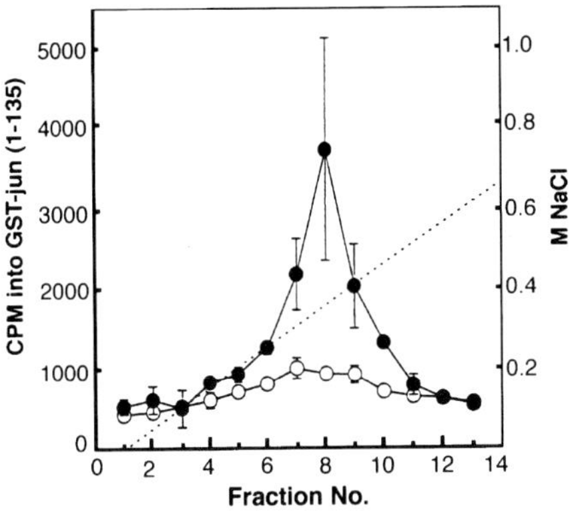

Fig. 3. Cation-exchange chromatography of the hepatic activator of JNK/SAPK. Liver cytosol from control (○) or IL-1-stimulated (●) rabbits was chromatographed on Fast Flow S Sepharose at pH 6.0 and eluted with a salt gradient. Fractions were assayed for the activator of JNK/SAPK. Reaction mixtures were separated by SDS/PAGE and phosphorylation of GST–Jun-(1–135) was measured by Čerenkov counting (c.p.m.). See reference for details [13]. Error bars show S.E.M. Reproduced from [13], Copyright 1997, with permission from Elsevier Science.

a protein band on electrophoresis, since the best preparations still contained a number of proteins [13].

The first JNK/SAPK activator to be identified was MKK4. It was isolated as a cDNA on the basis of its similarity with *Xenopus* MKK2 [25]. When the cDNA was transfected into mammalian cells, it activated JNK/SAPK and p38 [26,27]. Transfected MEKK1 (where MEK stands for MAP-kinase/ERK kinase, and MEKK is MEK kinase; see Fig. 1) was shown to activate transfected MKK4. However, although transfected MEKK1 activated endogenous MKK4, p38 was not activated, suggesting that, although MKK4 was biochemically capable of activating p38, physiologically it was selective for JNK/SAPK [26,27]. We found that the hepatic JNK/SAPK activator activated by IL-1 did not react with antibodies to MKK4; nor did it activate p38 *in vitro*. However, it did interact with antibodies made to a newly discovered activator, MKK7 [28,29]. MKK7 was identified as a mammalian homologue of the *Drosophila hemipterous* gene product, which is an activator of a *Drosophila* c-Jun kinase called *basket*. Null function mutants of these enzymes prevent the dorsal closure of the embryo, and the *hemipterous* null mutant was rescued by MKK7 [29]. The hepatic activator stimulated by IL-1 was immunoprecipitated by two different antisera to MKK7, one against the C-terminal half of the protein and one against the C-terminal peptide (Fig. 4). While this does not prove that the IL-1-induced activator is MKK7, it must be closely related. In view of the difficulty in purifying the activator to homogeneity for amino acid sequencing, the immunological identification must suffice until other MKK7 forms are discovered by cDNA cloning. During the course of our work, MKK7 was also identified by Lawler et al. [30] and shown to be activated by IL-1 in KB cells.

Upstream of MKKs

What lies upstream of MKK7 and MKK6? A number of MKK kinases (MKKKs) have been identified by cDNA cloning of homologues of c-Raf and related proteins (e.g. yeast Ste11) and by overexpression of the putative MKKKs in mammalian cells. These include MEKK1 [31], MAPK kinase kinase 5 (MAPKKK5) [32], apoptosis-stimulating kinase 1 (ASK1) [33] and transforming growth factor-β-activated kinase 1 (TAK1) [34]. The mixed-lineage kinases MLK3 [35] and dual leucine-zipper kinase [36] may also act as MKKKs. It is not clear how specific these enzymes are for activation of JNK/SAPK, since many of them also cause activation

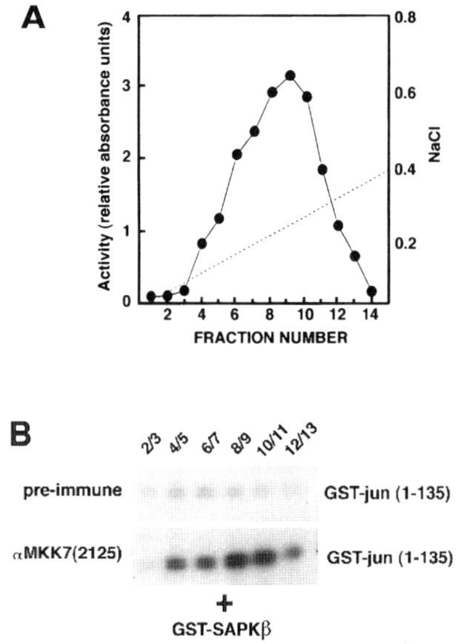

Fig. 4. Purification of the JNK/SAPK activator. (A) Chromatography of the JNK/SAPK activator on S Resource. Activator purified through Fast Flow S Sepharose (Fig. 3), phenyl-Sepharose and Q Resource chromatography was applied to an S Resource column, which was eluted with a salt gradient. Fractions were assayed for the activator of JNK/SAPK, as described for Fig. 3. Absorbance units were obtained by densitometry of an autoradiograph. (B) Antiserum 2125 to MKK7 C-terminal peptide immunoprecipitates the JNK/SAPK activator. Pairs of active fractions (numbers given above lanes) from (A) were treated with Protein A–agarose coated with antiserum or pre-immune serum, and the beads were incubated with GST–SAPKβ (25 μg/ml) in buffer containing ATP for 1 h at room temperature. GST–Jun and [γ-^{32}P]ATP were then added for a further 20 min. Phosphorylation of GST–Jun-(1–135) was detected by autoradiography after SDS/PAGE. Reproduced from [13], Copyright 1997, with permission from Elsevier Science.

of p38 when overexpressed in cells. Which, if any, of them mediate the effects of IL-1 (or TNF) remains to be established.

Precisely how the cytokine receptors activate the MAPK cascades is not known, but rapid progress has been made in the last few years in the discovery of molecules that interact with IL-1 and TNF receptors. These are summarized in Figs. 1(A) and 1(B). IL-1 (α and β behave identically) signals via its type 1 receptor, while both the inflammatory and cytotoxic actions of TNF are mediated by the p55 TNF receptor. Upon interaction with TNF, which is a trimer, the receptor dimerizes or trimerizes. It contains a death domain, which interacts with other death-domain-containing proteins such as TNF-receptor-associated death domain (TRADD) and Fas-associated death domain (FADD). These in turn bind TNF-receptor-associated factors (TRAFs) and a kinase called receptor-interacting protein (RIP) [37,38]. Overexpression of TRAF2 activates the JNK/SAPK and NF-κB pathways [39–41]. Another TRAF, TRAF6, has been implicated in IL-1 receptor signalling [42]. The IL-1 type 1 receptor interacts with an accessory protein and the IL-1-receptor-associated kinase upon ligand binding [43,44]. This complex may interact with TRAF6. Whether the TRAFs interact with MKKKs remains to be seen. This appears to be the case for the NF-κB pathway, because the IκB kinase, CHUK, is activated by an NF-κB-activating kinase [10], which is a homologue of the MKKKs and interacts with TRAF2 [45]. Thus NF-κB signalling and MAPK signalling seem to diverge below the level of TRAFs.

How interaction with TRAF2 activates NF-κB-activating kinase is unknown. TRAF2 might serve to bring the kinase into association with other receptor-associated kinases, or the proximity of NF-κB-activating kinase to the membrane may cause activation by other kinases, perhaps in a manner analogous to that undergone by Raf.

The MAPK and IκB kinase cascades are involved in IL-1 and TNF generation as well as in their action

The three MAPK cascades and the IκB kinase are strongly activated in monocytes and macrophages by microbial products such as bacterial LPS [46–48]. These signalling pathways are therefore implicated in the production of cytokines as well as in their action. TNF and IL-1, like most cytokines and many proteins made in inflammatory and immune responses, contain NF-κB sites in their promoters. Production of such proteins can be inhibited either by proteaseome inhibitors [49], which prevent degradation of IκB, or by overexpression of IκB, in order to prevent activation of NF-κB [50].

The importance of the MAPK pathways for cytokine production has still to be established. Inhibition of the activation of p42/p44 MAPKs in monocytes by means of the compound PD98059, which prevents MKK1/2 activation, has relatively small effects on cytokine production [51]. Inhibition of the p38 MAPK with the compound SB203580 does, however, inhibit IL-1 and TNF production in LPS-stimulated monocytes [52]. The basis of this inhibition is discussed below.

As yet there are no specific JNK/SAPK inhibitors available, so the importance of this pathway for cytokine production is not known. AP-1 sites are common in promoters of inflammatory genes, and JNK/SAPK is strongly implicated in regulating AP-1-binding proteins such as c-Jun and ATF2 [53].

The extent to which these pathways are involved in the generation of IL-1 and TNF in monocytes and macrophages in inflammatory diseases is not established, since our understanding of their production is almost entirely based on experiments with LPS. However, there are very strong indications that interfering with them could inhibit both the production and actions of IL-1 and TNF, so they provide many (perhaps too many) potential targets for therapy. Nevertheless, more needs to be known about the processes controlled by the pathways, both in normal physiology and in chronic inflammatory disease, before a particular pathway or kinase is selected as a target for anti-inflammatory drugs.

Downstream processes regulated by MAPKs

Experiments with the inhibitor of MKK1/2 activation (PD98059) suggest that the p42/p44 MAPKs do not play a major role either in the production of IL-1 or TNF [51] or in their pro-inflammatory actions. Furthermore, since this pathway is implicated in cell proliferation, one would intuitively avoid inhibiting it. The consequences of blocking the JNK/SAPK pathway are unknown, as mentioned above, because there are no small specific inhibitors. The known substrates of JNK/SAPK are transcription factors involved in the formation of AP-1-binding complexes, ternary complex factors such as Elk1, and the epidermal growth factor receptor (phosphorylation of Thr-669) [53].

In contrast with JNK/SAPK, much more is known about the consequences of blocking p38. The pyridinyl imidazole inhibitors discovered by Lee and Young at SmithKline Beecham have proved valuable for probing p38 function [52,54]. The best characterized is SB203580. It inhibits p38α and p38β (but not γ or δ) in the 0.1–1.0 μM concentration range. It does not affect the other MAPKs or many kinases tested. Like any inhibitor, its effects on cells need to be interpreted cautiously, but if it has a strong biological action in the 0.1–1.0 μM range this is very likely to be due to blockade of p38. Effects at higher concentrations may well be due to the compound interacting with other cellular systems.

We have investigated the effects of p38 blockade on the actions of IL-1, and found highly selective inhibition of the induction of cyclo-oxygenase 2 and of the matrix metalloproteinases collagenase and stromelysin (Table 1) [55]. IL-6 production induced by IL-1 was partially and variably inhibited (generally 30–40%), while IL-8 seemed unaffected. The inhibition of cyclo-oxygenase 2 and of the matrix metalloproteinases was

Table 1. Effects of SB203580 on IL-1 responses. Abbreviation: HUVEC, human umbilical vein endothelial cells.

IL-1-stimulated effect	Cell system	Time	Inhibition (%)
IL-6 production	Fibroblasts	16 h	~40
	HUVEC	16 h	~40
IL-8 production	Fibroblasts	16 h	0
	HUVEC	16 h	0
Prostaglandin E_2 production	Fibroblasts	16 h	95
	HuVEC	16 h	95
Cyclo-oxygenase 2 induction	Fibroblasts	16 h	95
Collagenase production	Fibroblasts	16 h	90
Stromelysin production	Fibroblasts	16 h	90
Proteoglycan degradation	Cartilage	24 h	0
Collagen degradation	Cartilage	14 days	75

at the level of mRNA accumulation. This was unexpected, since in other studies the pyridinyl imidazole inhibitors had been found to inhibit production of TNF and IL-1 by LPS-activated monocytes largely at the translational level [56]. The effect of the inhibitors on the induction of mRNA for the cytokines was small compared with their inhibition of production of cytokine protein. A subsequent more detailed study on the induction of TNF by LPS in the cell line THP-1 supported this earlier interpretation [57].

We have re-examined this question in elutriated human monocytes and measured the effects of the p38 inhibitor on TNFα and IL-1β production at both the mRNA and protein levels. The results are summarized in Table 2. We chose to use SB203580 at 2 μM, since this concentration inhibits intracellular p38 by more than 90% (as judged by inhibition of intracellular Hsp27 phosphorylation), and higher concentrations may have non-specific effects. Inhibition of LPS-induced cyclo-oxygenase 2 (about 80%) was maximal with this concentration. TNF inhibition was incomplete (about 65% on average at the protein level), and there was an almost equivalent partial inhibition of the steady-state level of mRNA. Production of IL-1β protein, on the other hand, was strongly inhibited, with a relatively small effect on the steady-state level of its mRNA. Thus it can

Table 2. Effects of the p38 MAPK inhibitor SB203580 (2 μM) on gene expression induced by bacterial LPS in primary human monocytes.

Protein	Stimulation time (h)	Protein inhibition (%)	Steady-state mRNA inhibition (%)
TNF	4	62	54
IL-1β	4	> 90	30
Cyclo-oxygenase 2	16	80	80

be concluded that cyclo-oxygenase 2, metalloproteinases and TNFα are being inhibited at a pre-translational level, whereas the inhibition of IL-1β may be translational.

Whether the pre-translational inhibition is at the level of transcription or is due to instability of mRNA is uncertain. IL-1 has been shown to have effects on both the transcription and stability of cyclo-oxygenase 2 [58] and collagenase [59] mRNAs. Collagenase is an AP-1-regulated gene [60], and prior induction of c-*jun* by IL-1 may be necessary for its activation. SB203580 did not inhibit IL-1 induction of c-*jun* in fibroblasts (results not shown).

The effects of p38 blockade suggest that there is still much to be learnt about its downstream targets. It phosphorylates MAP-kinase-activated protein kinase 2 (MAPKAPK-2), whose only known substrates so far are Hsp27 and the cAMP response element binding protein (CREB) [61] (see Fig. 1). p38 also phosphorylates a related kinase called MAPK integrating kinase (MnK) that phosphorylates eukaryotic initiation factor 4E in a regulatory manner. This phosphorylation may be important in the mechanism of translation [62,63]. MnK is also activated by p42/p44 MAPK, hence its name. p38 may also phosphorylate the transcription factors myocyte-enhancing factor [64], generally regarded as specific for muscle proteins, and cAMP response element-binding protein homologous protein [65], a transcription factor induced by irradiation damage to DNA which is related to the CCAAT-enhancer-binding protein group. p38β (or p38-2), which is widely expressed and activated by IL-1, has a slightly different substrate specificity. It is more active towards the transcription factors ATF2 and serum-response factor accessory protein 1 than is p38α, but it is inhibitable by SB203580 [14,15]. It is not clear from these putative targets how p38 is regulating cyclo-oxygenase and metalloproteinase expression.

Conclusion

There has been rapid progress in our understanding of protein kinase cascades involved in IL-1 and TNF signalling, and of the receptor-associated proteins that assemble upon receptor activation. The gaps in our knowledge between the receptor–protein complexes and the downstream kinase cascades will soon be filled in. Future challenges lie in understanding the processes controlled by these pathways in physiology and pathology and in explaining the significance of why there are so many closely related enzymes at all levels of the MAPK pathways.

We thank the Medical Research Council and the Arthritis and Rheumatism Council for their support.

References

1. Dinarello, C.A. (1996) Blood **87**, 2095–2147
2. Beutler, B. (1992) Tumor Necrosis Factors: the molecules and their emerging role in medicine, Raven Press, New York

3. Guesdon, F., Freshney, N., Waller, R.J., Rawlinson, L. and Saklatvala, J. (1993) J. Biol. Chem. **268**, 4236–4243
4. Kyriakis, J.M., Banerjee, P., Nikolakaki, E., Dai, T., Rubie, E.A., Ahmad, M.F., Avruch, J. and Woodgett, J.R. (1994) Nature (London) **369**, 156–160
5. Freshney, N.W., Rawlinson, L., Guesdon, F., Jones, E., Cowley, S., Hsuan, J. and Saklatvala, J. (1994) Cell **78**, 1039–1049
6. Raingeaud, J., Gupta, S., Rogers, J.S., Dickens, M., Han, J., Ulevitch, R.J. and Davis, R.J. (1995) J. Biol. Chem. **270**, 7420–7426
7. Kracht, M., Truong, O., Totty, N.F., Shiroo, M. and Saklatvala, J. (1994) J. Exp. Med. **180**, 2017–2025
8. Kracht, M., Shiroo, M., Marshall, C.J., Hsuan, J.J. and Saklatvala, J. (1994) Biochem. J. **302**, 897–905
9. DiDonato, J.A., Hayakawa, M., Rothwarf, D.M., Zandi, E. and Karin, M. (1997) Nature (London) **388**, 548–554
10. Regnier, C.H., Yeong Song, H., Gao, X., Goeddel, D.V., Cao, Z. and Rothe, M. (1997) Cell **90**, 373–383
11. Guesdon, F., Waller, R.J. and Saklatvala, J. (1994) Biochem. J. **304**, 761–768
12. Guesdon, F., Knight, G., Rawlinson, L.M. and Saklatvala, J. (1997) J. Biol. Chem. **272**, 30017–30024
13. Finch, A., Holland, P., Cooper, J., Saklatvala, J. and Kracht, M. (1997) FEBS Lett. **418**, 144–148
14. Jiang, Y., Chen, C., Li, Z., Guo, W., Gegner, J.A., Lin, S. and Han, J. (1996) J. Biol. Chem. **271**, 17920–17926
15. Stein, B., Yang, M.X., Young, D.B., Janknecht, R., Hunter, T., Murray, B.W. and Barbosa, M.S. (1997) J. Biol. Chem. **272**, 19509–19517
16. Mertens, S., Craxton, M. and Goedert, M. (1996) FEBS Lett. **383**, 273–276
17. Goedert, M., Cuenda, A., Craxton, M., Jakes, R. and Cohen, P. (1997) EMBO J. **16**, 3563–3571
18. Kumar, S., McDonnell, P.C., Gum, R.J., Hand, A.T., Lee, J.C. and Young, P.R. (1997) Biochem. Biophys. Res. Commun. **235**, 533–538
19. Saklatvala, J., Rawlinson, L.M., Marshall, C.J. and Kracht, M. (1993) FEBS Lett. **334**, 189–192
20. Saklatvala, J., Davis, W. and Guesdon, F. (1996) Philos. Trans. R. Soc. London B **351**, 151–157
21. Belka, C., Wiegmann, K., Adam, D., Holland, R., Neuloh, M., Herrmann, F., Kronke, M. and Brach, M.A. (1995) EMBO J. **14**, 1156–1165
22. Cuenda, A., Cohen, P., Buee-Scherrer, V. and Goedert, M. (1997) EMBO J. **16**, 295–305
23. Cuenda, A., Alonso, G., Morrice, N., Jones, M., Meier, R., Cohen, P. and Nebreda, A.R. (1996) EMBO J. **15**, 4156–4164
24. Han, J., Lee, J.D., Jian, Y., Li, Z., Feng, L. and Ulevitch, R.J. (1996) J. Biol. Chem. **271**, 2886–2891
25. Sanchez, I., Hughes, R.T., Mayer, B.J., Yee, K., Woodgett, J.R., Avruch, J., Kyriakis, J.M. and Zon, L.I. (1994) Nature (London) **372**, 794–798
26. Derijard, B., Raingeaud, J., Barrett, T., Wu, I.H., Han, J., Ulevitch, R.J. and Davis, R.J. (1995) Science **267**, 682–685
27. Lin, A., Minden, A., Martinetto, H., Claret, F.X., Lange-Carter, C., Mercurio, F., Johnson, G.L. and Karin, M. (1995) Science **268**, 286–290
28. Tournier, C., Whitmarsh, A.J., Cavanagh, J., Barrett, T. and Davis, R.J. (1997) Proc. Natl. Acad. Sci. U.S.A. **94**, 7337–7342

29. Holland, P.M., Magali, S., Campbell, J.S., Nosell, S. and Cooper, J.A. (1997) J. Biol. Chem. **272**, 24994–24998
30. Lawler, S., Cuenda, A., Goedert, M. and Cohen, P. (1997) FEBS Lett. **414**, 153–158
31. Yan, M., Dai, T., Deak, J.C., Kyriakis, J.M., Zon, L.I., Woodgett, J.R. and Templeton, D.J. (1994) Nature (London) **372**, 798–800
32. Wang, X.H.S., Diener, K., Jannuzzi, D., Trollinger, D., Tan, T.H., Lichenstein, H., Zukowski, M. and Yao, Z.B. (1996) J. Biol. Chem. **271**, 31607–31611
33. Ichijo, H., Nishida, E., Irie, K., ten Dijke, P., Saitoh, M., Moriguchi, T., Takagi, M., Matsumoto, K., Miyazono, K. and Gotoh, Y. (1997) Science **275**, 90–94
34. Yamaguchi, K., Shirakabe, K., Shibuya, H., Irie, K., Oishi, N., Taniguchi, T., Nishida, E. and Matsumoto, K. (1995) Science **270**, 2008–2011
35. Tibbles, L.A., Ing, Y.L., Kiefer, F., Chan, J., Iscove, N., Woodgett, J.R. and Lassam, N.J. (1996) EMBO J. **15**, 7026–7035
36. Fan, G., Merritt, S.E., Kortenjann, M., Shaw, P.E. and Holzman, L.B. (1996) J. Biol. Chem. **271**, 24788–24793
37. Hsu, H.L., Shu, H.B., Pan, M.G. and Goeddel, D.V. (1996) Cell **84**, 299–308
38. Hsu, H., Huang, J., Shu, H.B., Baichwal, V. and Goeddel, D.V. (1996) Immunity **4**, 387–396
39. Rothe, M., Sarma, V., Dixit, V.M. and Goeddel, D.V. (1995) Science **269**, 1424–1427
40. Natoli, G., Costanzo, A., Ianni, A., Templeton, D.J., Woodgett, J.R., Balsano, C. and Levrero, M. (1997) Science **275**, 200–203
41. Song, H.Y., Regnier, C.H., Kirschning, C.J., Ayres, T.M., Goeddel, D.V. and Rothe, M. (1997) Proc. Natl. Acad. Sci. U.S.A. **94**, 9792–9796
42. Cao, Z., Xiong, J., Takeuchi, M., Kurama, T. and Goeddel, D.V. (1996) Nature (London) **383**, 443–446
43. Cao, Z.D., Henzel, W.J. and Gao, X.O. (1996) Science **271**, 1128–1131
44. Wesche, H., Korherr, C., Kracht, M., Falk, W., Resch, K. and Martin, M.U. (1997) J. Biol. Chem. **272**, 7727–7731
45. Malinin, N.L., Boldin, M.P., Kovalenko, A.V. and Wallach, D. (1997) Nature (London) **385**, 540–544
46. Liu, M.K., Herrera, V.P., Brownsey, R.W. and Reiner, N.E. (1994) J. Immunol. **153**, 2642–2652
47. Hambleton, J., Weinstein, S.L., Lem, L. and DeFranco, A.L. (1996) Proc. Natl. Acad. Sci. U.S.A. **93**, 2774–2778
48. Han, J., Lee, J.D., Bibbs, L. and Ulevitch, R.J. (1994) Science **265**, 808–811
49. Read, M.A., Neish, A.S., Luscinskas, F.W., Palombella, V.J., Maniatis, T. and Collins, T. (1995) Immunity **2**, 493–506
50. Wrighton, C.J., Hofer-Warbinek, R., Moll, T., Eytner, R., Bach, F.H. and de Martin, R. (1996) J. Exp. Med. **183**, 1013–1022
51. Foey, A.D., Parry, S.L., Williams, L.M., Feldmann, M., Foxwell, B.J. and Brennan, F.M. (1998) J. Immunol. **160**, 920–928
52. Lee, J.C., Laydon, J.T., McDonnell, P.C., Gallagher, T.F., Kumar, S., Green, D., McNulty, D., Blumenthal, M.J., Heys, J.R., Landvatter, S.W., et al. (1994) Nature (London) **372**, 739–746
53. Karin, M., Liu, Z. and Zandi, M. (1997) Curr. Opin. Cell Biol. **9**, 240–246
54. Cuenda, A., Rouse, J., Doza, Y.N., Meier, R., Cohen, P., Gallagher, T.F., Young, P.R. and Lee, J.C. (1995) FEBS Lett. **364**, 229–233
55. Ridley, S.H., Sarsfield, S.J., Lee, J.C., Bigg, H.F., Cawston, T.E., Taylor, D.J., DeWitt, D.L. and Saklatvala, J. (1997) J. Immunol. **158**, 3165–3173
56. Young, P., McDonnell, P., Dunnington, D., Hand, A., Laydon, J. and Lee, J. (1993) Agents Actions **39**, C67–C69

57. Prichett, W., Hand, A., Sheilds, J. and Dunnington, D. (1995) J. Inflamm. **45**, 97–105
58. Ristimaki, A., Garfinkel, S., Wessendorf, J., Maciag, T. and Hla, T. (1994) J. Biol. Chem. **269**, 11769–11775
59. Vincenti, M.P., Coon, C.I., Lee, O. and Brinckerhoff, C.E. (1994) Nucleic Acids Res. **22**, 4818–4827
60. Auble, D.T. and Brinckerhoff, C.E. (1991) Biochemistry **30**, 4629–4635
61. Tan, Y., Rouse, J., Zhang, A., Cariati, S., Cohen, P. and Comb, M.J. (1996) EMBO J. **15**, 4629–4642
62. Waskiewicz, A.J., Flynn, A., Proud, C.G. and Cooper, J.A. (1997) EMBO J. **16**, 1909–1920
63. Fukunaga, R. and Hunter, T. (1997) EMBO J. **16**, 1921–1933
64. Han, J., Jiang, Y., Li, Z., Kravchenko, V.V. and Ulevitch, R.J. (1997) Nature (London) **386**, 296–299
65. Wang, X.Z. and Ron, D. (1996) Science **272**, 1347–1349

Biochem. Soc. Symp. **64**, 79–89
Printed in Great Britain

6

Regulation of actin dynamics by stress-activated protein kinase 2 (SAPK2)-dependent phosphorylation of heat-shock protein of 27 kDa (Hsp27)

Jacques Landry* and Jacques Huot

Centre de recherche en cancérologie de l'Université Laval,
Centre Hospitalier Universitaire de Québec, Pavillon HDQ,
11 Côte du Palais, Québec (PQ), Canada G1R 2J6

Abstract

Activation of the mitogen-activated protein kinase (MAP kinase) SAPK2 (stress-activated protein kinase 2) leads to the phosphorylation of several transcription factors and cytoplasmic proteins, and thereby presumably orchestrates important specific cellular responses to numerous cytokines, stressing agents and agonists of tyrosine kinase or serpentine receptors. The heat-shock protein of 27 kDa (Hsp27), a downstream target of the SAPK2-activated MAP-kinase-activated protein kinase-2/3, has a documented function as an actin polymerization modulator. Accordingly, recent evidence implicates the SAPK2 pathway in the modulation of microfilament dynamics in response to stress and agonist stimulation. In vascular endothelial cells, where the basal level of expression of Hsp27 is high, SAPK2 mediates oxidative stress- and vascular endothelial growth factor (VEGF)-induced actin reorganization and VEGF-induced cell migration, suggesting a key role for this MAP kinase pathway in inflammation and angiogenic processes.

Introduction

SAPK2 (stress-activated protein kinase 2; also called p38α/β or RK) is a member of the still expanding family of mitogen-activated protein

*To whom correspondence should be addressed.

kinases (MAP kinases) which have vital roles in transducing and orchestrating messages generated by a variety of growth factors, inflammatory cytokines and stressing agents [1–4]. Whereas the first group of MAP kinases characterized, represented by extracellular-signal-regulated kinases 1/2 (ERK1/2), have been shown mostly to be growth factor- or mitogen-activated kinases, SAPK2 (together with three other SAPKs) is mostly recognized as a stress-sensitive kinase. SAPK2 activation leads directly or indirectly via SAPK2-activated kinases to phosphorylation and activation of a number of transcription factors, including CCAAT-enhancer-binding protein homologous protein, cAMP response element binding protein (CREB) and Elk-1, and cytoplasmic proteins such as eukaryotic initiation factor 4E and Hsp27 (heat-shock protein of 27 kDa) [4,5]. During stress, SAPK2 may regulate the expression of a number of genes which, like c-*jun* and c-*fos*, influence growth or expression of differentiated functions, or may eventually lead to expression or modification of the activity of proteins that confer protection and increased repair capacity in response to stress. Hsp27 has been shown to be an important determinant of the resistance of cells to numerous cytotoxic agents [6]. Alternatively, activation of SAPK2 may participate in the induction of cell death. A balance between ERK1/2 and SAPK activities was suggested to be a determining factor in the induction of apoptosis [7].

The list of activators of SAPK2 extends well beyond stressing agents and includes a number of inflammatory cytokines and, in some cell systems, agonists of tyrosine kinase receptors and serpentine receptors. Based on the effects of the inhibitors SB203580 and SB202190, a number of cellular responses have been attributed to SAPK2 induction. In monocytes, activation of SAPK2 modulates the production of inflammatory cytokines in response to bacterial lipopolysaccharide endotoxin, and in murine L929 cells it mediates the tumour necrosis factor α-induced expression of interleukin-6 (IL-6) [8,9]. SAPK2 activation is also involved in the regulation of glucose uptake by IL-1 in epithelial cells, [10] and in the collagen-induced aggregation of platelets [11]. As reviewed here, stimulation of SAPK2 in vascular endothelial cells mediates F-actin reorganization and cell migration induced by vascular endothelial growth factor (VEGF) [12], and major cytoskeletal responses in cells exposed to oxidative stress [13]. A precise mechanism involving the SAPK2-dependent activation of MAP-kinase-activated protein kinase 2/3 (MAPKAPK-2/3), which itself phosphorylates Hsp27 [14–17], is suggested. It is proposed that the SAPK2–MAPKAPK-2/3–Hsp27 module represents a major signalling pathway responsible for modulating actin dynamics in response to physiological agonists and stressing agents.

Hsp27

Hsp27 (also called Hsp25) is a member of the family of small heat-shock proteins, which includes (in mammalian cells) p20 and αA- and αB-crystallins [6,18]. All proteins of this family share a conserved α-crystallin domain which is responsible for the ability of the protein to form large oligomers. Hsp27 is expressed at varying levels during development and in different adult tissues and cell lines. It is barely detectable in cell lines such as fibroblasts, whereas it is present at concentrations close to 1% of total protein in endothelial cells, muscle cells and some tumour cell lines [6]. As with other heat-shock proteins, the gene coding for Hsp27 is strongly activated by heat shock and other stresses. Ectopic expression of Hsp27 was shown to confer a high level of cellular resistance to heat shock, tumour necrosis factor, oxidative stress and a number of cytotoxic drugs such as doxorubicin and cisplatin [19–25]. Changes in the cellular redox state and inhibition of apoptosis induction have been described in some cell lines expressing ectopic Hsp27 [25,26].

Hsp27 is serine-phosphorylated at two (rodent) or three (human) sites [27,28]. Numerous agents induce the phosphorylation of Hsp27. These include stressing agents such as heat shock, oxidative stress and osmotic stress, inflammatory cytokines such as tumour necrosis factor α and IL-1β, and also growth/differentiation factors that use either seven-transmembrane serpentine receptors (thrombin, bombesin) or tyrosine kinase receptors {VEGF [21], fibroblast growth factor (FGF) and platelet-derived growth factor [6]}. In all cases studied, phosphorylation is accompanied by a rapid increase in Hsp27 kinase activity which has been identified unequivocally as MAPKAPK-2/3, a physiological substrate of SAPK2 [14–17,29,30]. For typical cytotoxins, cytokines and agonists of tyrosine kinase or serpentine receptors, activation of SAPK2 accompanies and is required for Hsp27 phosphorylation. In all cases investigated, Hsp27 phosphorylation and MAPKAPK-2/3 activation are blocked by the SAPK2 inhibitor SB203580, whereas inhibiting ERK1/2 activation by expression of an interfering mutant of *ras* or treatment with PD098059 (an inhibitor of the ERK pathway) has no effect on Hsp27 phosphorylation, even during agonist stimulation [21,22,25].

Phosphorylation-modulated functions of Hsp27

Phosphorylation induces dramatic changes in the supramolecular structure of Hsp27. Size fractionation, ultracentrifugation and cross-linking experiments indicate that Hsp27 is organized, *in vivo*, into structures of varying sizes in equilibrium that consist of multimers of homodimers and also of monomers. The equilibrium is rapidly displaced towards small molecular species upon phosphorylation, indicating that phosphorylation destabilizes Hsp27–Hsp27 interactions. Wild-type Hsp27 in unstimulated cells or non-phosphorylatable mutants of Hsp27 form mainly 15 S (~16-mer) structures, whereas wild-type Hsp27 in stimulated cells or pseudo-phosphorylated mutants of Hsp27 are found mainly as dimers and monomers ([18,23]; H. Lambert and J. Landry, unpublished work).

Two documented biochemical activities of Hsp27 are potentially modulated by phosphorylation and/or phosphorylation-induced changes in the supramolecular structure of the protein. Recombinant Hsp27, which is mostly organized into 16-mer complexes, possesses chaperone properties [31]. In this state, Hsp27 can bind partially denatured proteins and thereby prevent their irreversible aggregation. It is postulated that its function is to trap denatured proteins in a folding-competent state, allowing subsequent refolding in co-operation with other chaperones such as Hsp70 [32,33]. Phosphorylation of the recombinant protein was reported to have no significant effect on the chaperone activity. However, in contrast with Hsp27 expressed in mammalian cells, the multimeric structure of recombinant Hsp27 is not affected by phosphorylation [34]. One may hypothesize that, *in situ*, the small dimeric or monomeric Hsp27 produced by phosphorylation would lack chaperone activity.

Purified Hsp27 also behaves *in vitro* as an F-actin capping protein, reducing the low-shear viscosity of F-actin solutions and inhibiting actin polymerization [35]. This activity is restricted to non-phosphorylated monomers of Hsp27 and thus appears to be distinct from the chaperone activity. Non-phosphorylated monomers inhibit actin polymerization by 90% at an actin/Hsp27 molar ratio of 1:1, whereas phosphorylated monomers, non-phosphorylated oligomers and recombinant Hsp27 are inactive in actin polymerization assays [36]. Direct evidence that Hsp27 also interacts with actin filaments *in vivo* was obtained by showing that cells overexpressing Hsp27 following gene transfection grow and survive better on exposure to cytochalasin D than do control cells [23]. Cytochalasin D reversibly binds the barbed end of actin filaments and inhibits polymerization. As a result, actin filaments in cells exposed to cytochalasin D are rapidly depolymerized. Depolymerization occurs less rapidly in cells that express high levels of Hsp27 compared with control cells; moreover, filaments re-polymerize more rapidly in these cells during recovery from cytochalasin D exposure [23,37]. Evidence was also obtained *in vivo* that the activity of Hsp27 is dependent on its state of phosphorylation and size. Overexpression of a non-phosphorylatable mutant of Hsp27 which was mostly organized as high-molecular-mass multimers had no effect on actin dynamics, and did not prevent cytochalasin D-induced growth inhibition. Furthermore, treatment of the cells with the SAPK2 inhibitor SB203580, which results in decreased phosphorylation of Hsp27 and a shift of Hsp27 towards the high-molecular-mass structures, totally inhibited Hsp27-dependent protection against cytochalasin D-induced filament disruption. In contrast with SB203580, treatment with sodium arsenite, which induces phosphorylation of Hsp27 and increases the proportion of low-molecular-mass Hsp27, potentiates the protective effect of Hsp27 [17].

Actin filaments are highly sensitive to the presence of cytotoxic agents. Alterations in structures, depolymerization, loss of stress fibres and formation of F-actin patches are common features after treatments such as heat shock, oxidative stress or hyperosmotic stress. For all toxic agents investigated, cells overexpressing Hsp27 were found to have an actin cytoskeleton that was more resistant to disruption, and/or that recovered more rapidly from disruption, than that in control cells or cells overexpressing a non-phosphorylatable mutant of Hsp27. This indicates that phosphorylation of Hsp27 confers resistance against stress-induced microfilament disorganization [21,22].

The SAPK2 pathway mediates actin reorganization in response to stress and growth factors

F-actin remodelling is one of the earliest responses induced in cells exposed to stress or growth factors. Regulation of actin polymerization and organization in mammalian cells is a highly complex process which involves a number of actin-binding proteins, including severing, sequestering, cross-linking and membrane-anchoring proteins. In this context of competition between the various effectors of actin dynamics, the importance of the SAPK2/Hsp27 pathway in regulating responses of the microfilaments is expected to be related to the cellular concentrations of Hsp27. In cells such as most fibroblasts, in which the amount of Hsp27 is very low, activation of the SAPK2 pathway is not expected to have much influence on the microfilament response. In contrast, in cells such as those of the vascular endothelium, skeletal muscle and kidney, and in most cell types after stress, high levels of Hsp27 expression may result in a major role for the SAPK2 pathway in modulating actin dynamics and actin-organization-dependent processes. The first evidence for a role for Hsp27 in modulating microfilament responses to external factors was obtained in fibroblastic cells in which the level of Hsp27 was artificially increased by gene transfection. Like several of the actin-binding proteins, Hsp27 has an enhanced localization in cell protrusions such as lamellipodia, filopodia and membrane ruffles. In fibroblastic cell lines that overexpress Hsp27, F-actin is more cortical, and pinocytotic activity is increased relative to control cells that express little Hsp27. In response to thrombin or FGF, actin polymerization is enhanced in these cells, whereas in cells that overexpress a non-phosphorylatable mutant of Hsp27 the actin response to growth factors is blocked [37].

The physiological importance of the SAPK2 pathway in modulating actin dynamics and actin-organization-dependent processes was investigated using human umbilical vein endothelial cells (HUVEC) [12,13]. These cells constitutively express high levels of Hsp27; also, because of their proximity to the bloodstream, they are continuously exposed *in vivo* to 'stressing' agents or conditions causing oxidative, osmotic or shear stresses. Furthermore, several functions of the endothelium are known to be dependent on the dynamics of the microfilaments, such as trans-endothelium permeability, retraction and migration during angiogenesis. In unstimulated primary cultures of HUVEC, F-actin is found mostly in cortical structures forming characteristic ruffles (Figs. 1A and 1B). Upon stimulation with VEGF (a mitogenic, motogenic and important angiogenic factor) or with H_2O_2 (a by-product of inflammation and second messenger of several agonists of endothelial cells), F-actin reorganizes within 10–15 min into long trans-cytoplasmic stress fibres (Figs. 1C and 1D). In both cases, stress-fibre formation is associated with the assembly of vinculin to focal adhesions (not shown; see [12,13]). Pretreatment of the cells with SB203580 inhibits, in a dose-dependent manner, both the formation of stress fibres (Figs. 1E and 1F) and the assembly of vinculin to focal adhesions. The inhibitory effects of SB203580 are first apparent at concentrations similar to those that inhibit the SAPK2-dependent activation of MAPKAPK-2/3 and Hsp27 phosphorylation induced by VEGF and H_2O_2 [12,13]. These results, taken together, suggest that the microfilament reorganization and assembly of focal adhesions triggered by VEGF and oxidative stress are mediated in HUVEC by activation of the SAPK2 pathway.

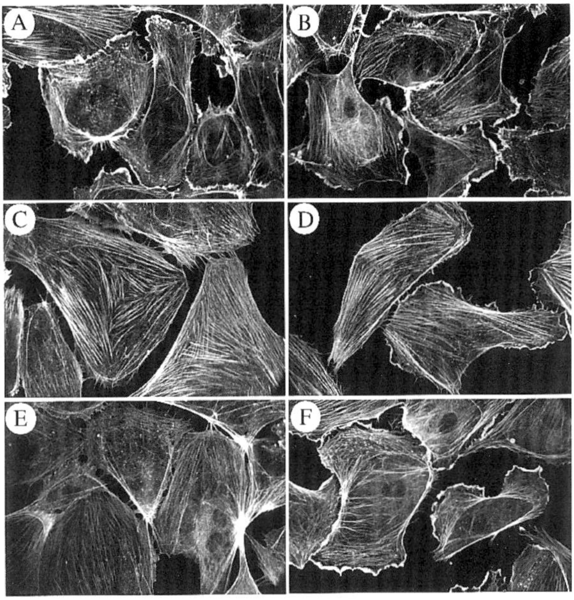

Fig. 1. H_2O_2 and the tyrosine kinase receptor agonist VEGF induce similar SB203580-sensitive reorganization of F-actin in primary cultures of endothelial cells. HUVEC were left untreated (A, B), or were treated with H_2O_2 (15 min, 250 μM) (C, E) or VEGF (15 min, 5 ng/ml) (D, F) in the absence (C, D) or presence of the SAPK2 inhibitor SB203580 (E, 5 μM; F, 1 μM). F-actin was stained with fluorescein-isothiocyanate-conjugated phalloidin and visualized by confocal microscopy using a $\times 60$ objective with a numerical aperture of 1.4 (scale, 1 cm = 50 μm).

Phosphorylation of Hsp27 mediates the SAPK2-dependent modulation of F-actin organization

The response of microfilaments to H_2O_2 in HUVEC contrasts with that in fibroblasts, where the oxidant produces a severe disorganization of the microfilament network characterized by fragmentation of F-actin, which forms patches concentrated around the nucleus (Figs. 2A and 2B). HUVEC express high levels of Hsp27 (approx. 6 ng/μg of total protein)

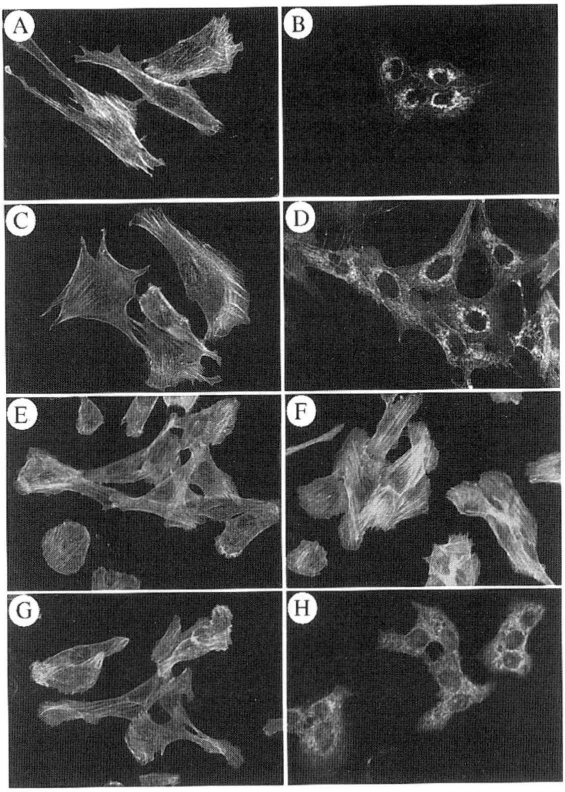

Fig. 2. Concentration- and SAPK2-induced phosphorylation of Hsp27 modulates the microfilament response to H_2O_2 in fibroblasts. Cells from the Chinese hamster cell line CCL39 clone 3 (A, B), clone V (C, D) and clone B12 (E–H) were left untreated (A, C, E, G) or treated for 1 h with 800 μM H_2O_2 (B, D, F, H). In (G) and (H), H_2O_2 treatment was carried out in the presence of the SAPK2 inhibitor SB203580 (25 μM). In addition to the basal level of Hsp27 (0.1% of total protein) found in the control cell line, clone V expresses a phosphorylation mutant of human Hsp27 at a concentration of 0.6% and clone B12 expresses wild-type human Hsp27 at a concentration of 0.6%. F-actin was stained with fluorescein-isothiocyanate-conjugated phalloidin and visualized by confocal microscopy using a $\times 60$ objective with a numerical aperture of 1.4 (scale, 1 cm = 50 μm).

compared with the low levels found in fibroblasts (often lower than 1 ng/μg of total protein). Interestingly, overexpression of Hsp27 in fibroblasts to levels equivalent to those found in HUVEC makes the fibroblasts respond to H_2O_2 in a way reminiscent of the response of HUVEC. Not only is F-actin in Hsp27-overexpressing fibroblasts protected against the treatment, but H_2O_2 induces in these cells, as in HUVEC, a rearrangement of microfilaments, which reorganize into long stress fibres (Figs. 2E and 2F). Overexpression of non-phosphorylatable Hsp27 is ineffective in modulating stress-fibre formation or preserving F-actin integrity in response to H_2O_2 (Figs. 2C and 2D). Furthermore, pretreating the cells with SB203580 totally abolished the effect of overexpressing wild-type Hsp27 (Figs. 2G and 2H). These results indicate that the SAPK2-dependent phosphorylation of Hsp27 is involved in maintaining the integrity of F-actin in the presence of H_2O_2 and in mediating the H_2O_2-induced shift from cortical microfilaments to stress fibres, both in fibroblasts artificially overexpressing Hsp27 and in HUVEC which naturally express high levels of Hsp27.

Cellular functions regulated by the SAPK2/Hsp27-dependent modulation of actin dynamics

VEGF is a major regulator of both physiological and pathological angiogenesis. It is an endothelial-cell-specific mitogenic and chemotactic agent produced by endothelial cells (endocrine pathway) and other cells such as smooth muscle and tumour cells (paracrine pathway). VEGF induces the production by endothelial cells of enzymes required for the degradation of extracellular-matrix proteins, changes in the permeability of the endothelium and the migration of endothelial cells, an essential process involved in neovascularization and requiring changes in microfilament dynamics [12,38]. Induction by VEGF of endothelial cell migration in culture occurs at the same concentrations that activate SAPK2 and actin reorganization [12]. Interestingly, as with F-actin reorganization, SB203580 drastically inhibits VEGF-induced cell migration (Fig. 3). Inhibition of activation of the ERK pathway has no effect on either actin reorganization or the induction of cell migration, but inhibits VEGF stimulation of DNA synthesis [12].

Concluding remarks

The microfilament network plays a key role in modulating cell shape, cell motility, cell division and muscle contraction. Moreover, in many cell types the integrity of the actin cytoskeleton is one of the earliest targets of stress, and it may well be a limiting factor of cellular resistance. The finding that the SAPK2/Hsp27 pathway is an important determinant of actin dynamics in response to growth factors or stress identifies this pathway as a key regulator of cellular responses triggered by environmental stimulation in both physiological and pathological situations. The role of the SAPK2 pathway in these processes can be predicted to be of importance, especially in cells that express, either naturally or after stress, high levels of Hsp27. Both major vascular cell types (endothelial and smooth-muscle cells), and many circulating cells, meet this requirement; furthermore, they have functions that are regulated by known potent activators of the SAPK2 pathway, including cytokines, active oxygen species and other stresses. In addition to the results presented here suggesting a role for SAPK2/Hsp27 in inflammation and angiogenic processes, a role for Hsp27 has been suggested in the bombesin-induced contraction of smooth muscle cells [39], in the release of FGF by endothelial cells [40] and in CD95 (APO-1/FAS)-mediated apoptosis [26]. In the light of these accumulating results, and considering that Hsp27 is only one of several targets of the SAPK2 pathways, future research in this area promises to be very revealing.

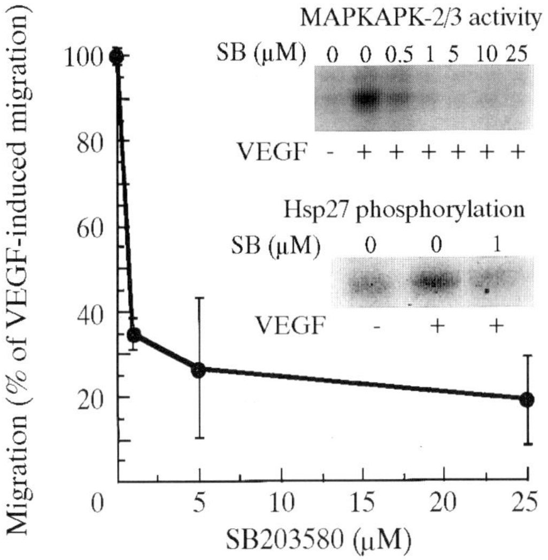

Fig. 3. The SAPK2 inhibitor SB203580 blocks VEGF-induced MAPKAPK-2/3 activation, Hsp27 phosphorylation and migration of HUVEC. The cells were plated on a 8 μm-pore-size gelatinized polycarbonate membrane, and migration across the membrane was induced by adding VEGF (5 ng/ml) for 2 h in the lower chamber, with varying concentrations of the inhibitor in the upper and lower chambers. Results are expressed relative to the number of cells that migrated in the absence of the inhibitor. The inset shows the effect of SB203580 on VEGF-induced MAPKAPK-2/3 activation and the *in vivo* phosphorylation of Hsp27. Adapted from [12] with permission.

The work of the authors reported here was supported by grants from the Medical Research Council of Canada and the Cancer Research Society Inc.

References

1. Waskiewicz, A.J. and Cooper, J.A. (1995) Curr. Opin. Cell Biol. **7**, 798–805
2. Treisman, R. (1996) Curr. Opin. Cell Biol. **8**, 205–215
3. Kyriakis, J. and Avruch, J. (1996) BioEssays **18**, 567–577
4. Goedert, M., Cuenda, A., Craxton, M., Jakes, R. and Cohen, P. (1997) EMBO J. **16**, 3563–3571
5. Cuenda, A., Cohen, P., Buée-Scherrer, V. and Goedert, M. (1997) EMBO J. **16**, 295–305
6. Arrigo, A.P. and Landry, J. (1994) in The Biology of Heat Shock Proteins and Molecular Chaperones (Morimoto, R.I., Tissière, A. and Georgopoulos, C., eds.), pp. 335–373, Cold Spring Harbor Laboratory Press, Cold Spring Harbor
7. Zhengui, X., Dickens, M., Raingeaud, J., Davis, R.J. and Greenberg, M.E. (1995) Science **270**, 1326–1331
8. Lee, J.C., Laydon, J.T., McConnell, P.C., Gallagher, T.F., Kumar, S., Green, D., McNulty, D., Blumenthal M.J., Heys, J.R., Landvatter, S.W., et al. (1994) Nature (London) **372**, 739–746
9. Beyaert, R., Cuenda, A., Berghe, W.W., Plaisance, S., Lee, J.C., Haegeman, G., Cohen, P. and Fiers, W. (1996) EMBO J. **15**, 1914–1923
10. Gould, G.W., Cuenda, A., Thomson, F.J. and Cohen, P. (1995) Biochem. J. **311**, 735–738
11. Saklatvala, J., Rawlinson, L., Waller, R.J., Sarsfield, S., Lee, J.C., Morton, L.F., Barnes, M.J. and Farndale, R.W. (1996) J. Biol. Chem. **271**, 6586–6589
12. Rousseau, S., Houle, F., Landry, J. and Huot, J. (1997) Oncogene **15**, 2169–2177
13. Huot, J., Houle, F., Marceau, F. and Landry, J. (1997) Circ. Res. **80**, 383–392
14. Cuenda, A., Rouse, J., Doza, Y.N., Meier, R. and Cohen, P. (1995) FEBS Lett. **364**, 229–233
15. Huot, J., Lambert, H., Lavoie, J.N., Guimond, A., Houle, F. and Landry, J. (1995) Eur. J. Biochem. **227**, 416–427
16. McLaughlin, M.M., Kumar, S., McDonnell, P.C, Van Horn, S., Lee, J.C., Livi, G.P. and Young, P.R. (1996) J. Biol Chem. **271**, 8488–8492
17. Guay, J., Lambert, H., Gingras-Breton, G., Lavoie, J.N., Huot, J. and Landry, J. (1997) J. Cell Sci. **110**, 357–368
18. Kato, K., Goto, S., Inaguma, Y., Hasegawa, K., Morishita, R. and Asano T. (1994) J. Biol. Chem. **269**, 1–9
19. Landry, J., Chrétien, P., Lambert, H., Hickey, E. and Weber, L.A. (1989) J. Cell Biol. **109**, 7–15
20. Huot, J., Roy, G., Lambert, H., Chrétien, P. and Landry, J. (1991) Cancer Res. **51**, 5245–5252
21. Huot, J., Houle, F., Spitz, D.R. and Landry, J. (1996) Cancer Res. **56**, 273–279
22. Lavoie, J.N., Gingras-Breton, G., Tanguay, R.M. and Landry, J. (1993) J. Biol. Chem. **268**, 24210–24214
23. Lavoie, J.N., Lambert, H., Hickey, E., Weber, L.A. and Landry, J. (1995) Mol. Cell. Biol. **15**, 505–516
24. Garrido, C., Ottavi, P., Fromentin, A. Hammann, A., Arrigo, A.-P., Chauffert, B. and Mehlen, P. (1997) Cancer Res. **57**, 2661–2667
25. Mehlen, P., Kretz-Remy, C., Préville, X. and Arrigo, A.P. (1996) EMBO J. **15**, 2695–2706

26. Mehlen, P., Schulze-Ostoff, K. and Arrigo, A.P. (1996) J. Biol. Chem. **271**, 16510–16514
27. Gaestel, M., Schröeder, W., Benndorf, R., Lippman, C., Buchner, K., Huchot, F., Erman, V.A. and Bielka, H. (1991) J. Biol. Chem. **266**, 14721–14724
28. Landry, J., Lambert, H., Zhou, M., Lavoie, J.N., Hickey, E., Weber, L.A. and Anderson, C.W. (1992) J. Biol. Chem. **267**, 794–803
29. Rouse, J., Cohen, P., Trigon, S., Morange, M., Alonso-Llamazares, A., Zamanillo, D., Hunt, T. and Nebrada, A.R. (1994) Cell **78**, 1027–1037
30. Freshney, N.W., Rawlinson, L., Guesdon, F., Jones, E., Cowley, S., Hsuan, J. and Saklatvala, J. (1994) Cell **78**, 1039–1049
31. Jakob, U., Gaestel, M., Engel, K. and Buchner, J. (1993) J. Biol. Chem. **268**, 1517–1520
32. Ehrnsperger, M., Gräber, S., Gaestel, M. and Buchner, J. (1997) EMBO J. **16**, 221–229
33. Lee, G.J., Roseman, A.M., Saibil, H.R. and Vierling, E. (1997) EMBO J. **16**, 659–671
34. Knauf, U., Jakob, U., Engel, K., Buchner, J. and Gaestel, M. (1994) EMBO J. **13**, 54–60
35. Miron, T., Wilcheck, M. and Geiger, B. (1988) Eur. J. Biochem. **178**, 543–553
36. Benndorf, R., Haye K., Ryazantsev, S., Wieske, M., Behlke, J. and Lutsch, G. (1994) J. Biol. Chem. **269**, 20780–20784
37. Lavoie, J.N., Hickey, E., Weber, L.A. and Landry, J. (1993) J. Biol. Chem. **268**, 24210–24214
38. Breier, G. and Risau, W. (1996) Trends Cell Biol. **6**, 454–456
39. Bitar, K.N., Kaminski, M.S., Hailat, N., Cease, K.B. and Strahler, J.R. (1991) Biochem. Biophys. Res. Commun. **181**, 1192–1200
40. Piotrowicz, R.S., Martin, J.L., Dillman, W.H. and Levin, E.G. (1997) J. Biol. Chem. **272**, 7042–7047

Biochem. Soc. Symp. **64**, 91–104
Printed in Great Britain

7

DNA-dependent protein kinase and related proteins

Graeme C.M. Smith*, Nullin Divecha†, Nicholas D. Lakin* and Stephen P. Jackson*‡

*Wellcome Trust/Cancer Research Campaign Institute of Cancer and Developmental Biology and Department of Zoology, University of Cambridge, Tennis Court Road, Cambridge CB2 1QR, U.K., and †Inositide Laboratory, Babraham Institute, Cambridge CB2 4AT, U.K.

Abstract

The DNA-dependent protein kinase (DNA-PK) is a nuclear protein serine/threonine kinase that must bind to DNA double-strand breaks to be active. We and others have shown that it is a multiprotein complex comprising an approx. 465 kDa catalytic subunit (DNA-PK_{cs}) and a DNA-binding component, Ku. Notably, cells defective in DNA-PK are hypersensitive to ionizing radiation. Thus X-ray-sensitive hamster *xrs-6* cells are mutated in Ku, and rodent V3 cells and cells of the severe combined immune-deficient (Scid) mouse lack a functional DNA-PK_{cs}. Cloning of the DNA-PK_{cs} cDNA revealed that it falls into the phosphatidylinositol (PI) 3-kinase family of proteins. However, biochemical assays indicate that DNA-PK contains no intrinsic lipid kinase activity, but is instead a serine/threonine kinase. We have also found that DNA-PK activity can be inhibited by the PI 3-kinase inhibitors wortmannin and LY294002. Consistent with its proposed role in genome surveillance and the detection of DNA damage, DNA-PK_{cs} is most similar to a subset of proteins involved in cell-cycle checkpoint control and signalling of DNA damage. Furthermore, the recent cloning of the gene mutated in ataxia–telangiectasia (A-T) patients, named *ATM* (A-T mutated), has revealed that the product of this gene is also a PI 3-kinase family member and is related to DNA-PK_{cs}. Although much is known about the clinical symptoms and cellular phenotypes that arise from disruption of the A-T gene, little is known about the biochemical action of ATM in response to DNA damage. Given its sequence similarity with DNA-PK_{cs}, we speculate that ATM may function in a manner similar to DNA-PK.

‡To whom correspondence should be addressed, at Wellcome/CRC Institute.

Introduction

The DNA-dependent protein kinase (DNA-PK) was first described as an activity in animal cell extracts that resulted in the phosphorylation of a variety of endogenous proteins, including the heat-shock protein Hsp90 [1]. Further studies revealed that DNA-PK must be bound to DNA before it can efficiently elicit its kinase activity [2–8]. Indeed, optimal substrates for DNA-PK are those that are DNA-binding proteins, suggesting that phosphorylation occurs most effectively when DNA-PK and its substrate are localized on the same DNA molecule. Initially, linearized double-stranded DNA molecules were thought to be the only activators of DNA-PK, although other types of DNA discontinuities in the DNA double helix have also been reported to activate its kinase potential in certain situations [9]. Phosphopeptide mapping and peptide substrate analyses showed that DNA-PK phophorylates proteins on serine and threonine residues, with a preference for the consensus sequence Ser/Thr-Gln [5,6,10].

The DNA-PK holoenzyme has been shown to be composed of two components: a DNA binding component, termed Ku, and a large $\sim$465 kDa catalytic subunit termed DNA-PK$_{cs}$ (for reviews see [11–13]). Ku, initially characterized as human autoimmune antigen, is a heterodimer composed of polypeptides of approx. 70 and 80 kDa (Ku70 and Ku80 respectively). This highly abundant protein [estimated at $(1–5) \times 10^5$ copies per human cell] binds tightly to double-stranded DNA ends and to other disruptions of the double-stranded DNA molecule. Ku thus binds to the DNA and, in so doing, recruits and activates the large catalytic subunit.

DNA-PK's ability to bind to and be activated by DNA led to a variety of proposals for a physiological role for this abundant and ubiquitously expressed kinase in mammalian cells (Fig. 1). A role for DNA-PK in modulating transcription has been put forward, since several transcription factors are phosphorylated by this kinase in a DNA-dependent manner *in vitro*. It has also been shown that DNA-PK can inhibit transcription mediated by RNA polymerase I *in vitro* [14,15]. In addition, the fact that DNA-PK binds non-sequence-specifically to DNA breaks via Ku led to the suggestion that DNA-PK may play a role *in vivo* in the repair of double-stranded DNA lesions induced by DNA-damaging agents, such as ionizing radiation. Indeed, it has been demonstrated over the last few years that certain cell lines which are hypersensitive to ionizing radiation and deficient in DNA double-strand-break rejoining are defective in components of DNA-PK (for reviews, see [16–18]). For example, the hamster cell line *xrs-6* is mutated in the Ku80 subunit, while the rodent V3 cell line and cells derived from the severe combined immune-deficient (Scid) mouse lack functional DNA-PK$_{cs}$. Another physiological role for DNA-PK, also demonstrated to be lacking in the aforementioned cell lines, is in the process of V(D)J recombination. This form of site-specific non-homologous genomic recombination is critical for the generation of the variable regions of immunoglobulin and T-cell receptor proteins. In developing B- and T-lymphocytes that lack DNA-PK function, such as those found in the Scid mouse, the V(D)J recombination intermediates are not processed or ligated effectively. This results in the Scid mouse lacking a functional immune system. Consistent with this, mice homozygous for a null mutation of Ku80 have a Scid phenotype [19].

Although cloning of the Ku subunits from a variety of eukaryotic species, ranging from yeast to humans, has taken place, sequence analysis has told us fairly little about the structural organization of this heterodimer at a protein level. However, both subunits have been reported to posses leucine-zipper motifs, suggesting that these regions of the proteins may be involved in dimerization or interaction with DNA-PK$_{cs}$. Recently, studies on the protein–protein and protein–DNA interactions within Ku have begun to take place ([20]; D. Gell and S.P. Jackson, unpublished work). Furthermore, development of yeast strains defective in either Ku70 or Ku80 has led to illuminating insights into the mode of action of this

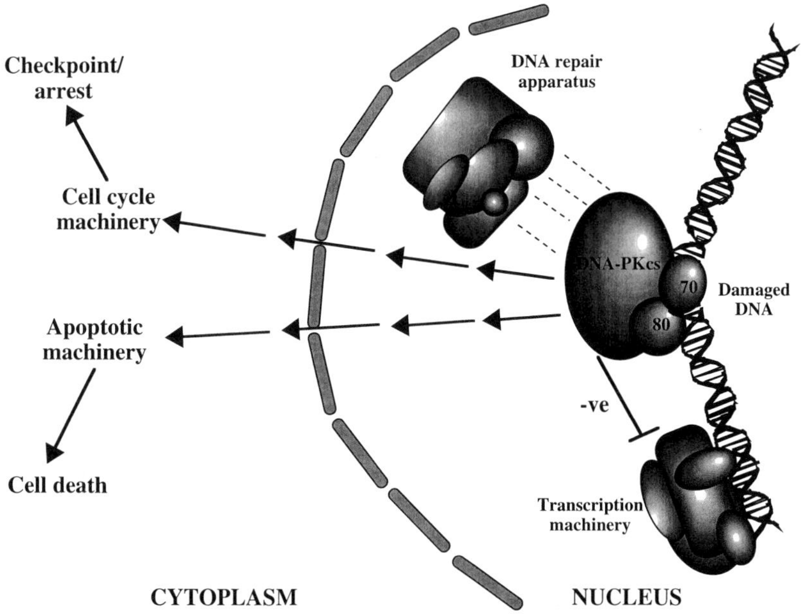

Fig. 1. Possible mechanism of DNA-PK function *in vivo* at a DNA break. At the site of a DNA break, DNA-PK$_{cs}$ is recruited to the DNA lesion via the non-sequence-specific DNA-binding protein Ku (Ku70/Ku80). This allows activation of the kinase function of the DNA-PK holoenzyme. This stimulation of activity may lead to inhibition of the cell-cycle machinery, inhibition of transcriptional activation and/or activation of the apoptotic machinery. Through protein–protein interactions, the DNA-PK holoenzyme may also act as a 'scaffold' at the DNA lesion to bring other components of the repair apparatus (e.g. ligases, endonucleases, polymerases) to the site of damage.

protein in double-strand-break repair and in the process of non-homologous end joining [21–25]. These mutant yeasts have also yielded significant findings indicating a role for Ku in telomeric length maintenance [22,26].

The cloning of DNA-PK$_{cs}$ led to the exciting discovery that the kinase domain of this protein is most similar to members of the phosphatidylinositol (PI) kinase family, but not to classical serine/threonine kinases [27]. This has led us to question whether DNA-PK has the capability of phosphorylating lipids. At around the same time as the DNA-PK$_{cs}$ cDNA was cloned, a range of other related large proteins involved in genomic surveillance and/or cell-cycle control were also identified (reviewed in [12,28]). One of these proteins, ATM ('A-T mutated'), is the protein that, when mutated, leads to the human disease state ataxia–telangiectasia (A-T) [29]. What we have learnt so far about DNA-PK may provide pointers to the biochemical functions of ATM and related proteins.

Structural analysis of DNA-PK$_{cs}$

The cloning of the cDNA of DNA-PK$_{cs}$ revealed an open reading frame that would encode a protein of approx. 465 kDa [27]. One of the most intriguing aspects of the derived amino acid sequence is that, surprisingly, the first 3500 amino acid residues do not contain any regions of significant identity with any other proteins in the databases. Furthermore, apart from a leucine-zipper motif situated from residues 1503 to 1538, no other motifs have been identified that could give a clear clue to the structure or function of the DNA-PK$_{cs}$ polypeptide. However, the C-terminal ~400 amino acid residues show high sequence similarity with the catalytic region of proteins that have been demonstrated to be PI kinases [30,31]; the p110 subunit of human PI 3-kinase, the yeast PI 3-kinase Vps34p and the human PI 4-kinase are shown in Fig. 2. At around the same time as the cloning of the DNA-PK$_{cs}$ cDNA was achieved, a variety of other proteins that had been implicated in DNA repair, genome surveillance and cell-cycle checkpoint controls were isolated by a host of groups working on a variety of different systems (reviewed in [12,28,32]). Sequence comparisons among these proteins revealed that they contained significant regions of identity in their C-terminal ~400 amino acids (Fig. 2). Members of this family include ATM [29] and its closest yeast homologue Tel1p [33,34], the *Saccharomyces cerevisiae* cell-cycle checkpoint regulatory protein Mec1p [35,36], its *Schizosaccharomyces pombe* homologue Rad3 [37,38] and its human homologue FRP (FRAP-related protein)/ATR (A-T-related) protein [39]. Related to this latter set of proteins is the *Drosophila* protein Mei-41 [40]. Also in this large group of proteins are the downstream targets for the immunosuppressant rapamycin, i.e. *S. cerevisiae* Tor1p (where Tor stands for target of rapamycin; see below) and the human protein FRAP

(FKBP12–rapamycin-associated protein, where FKBP12 is FK506-binding protein; see below) [41–46].

From Fig. 2, it can be seen clearly that all of the above proteins sharing a similar C-terminal kinase domain can be divided into two distinct groups. What has been termed subgroup 1 contains the smaller proteins, including p110, Vps34p and their PI 4-kinase PIK1. Subgroup 2 contains the larger members of the family, and includes ATM, Tel1p, Mec1p, Rad3, ATR, Mei-41, Tor1p, Tor2p and FRAP (see below for further descriptions of these proteins). A variety of criteria can be used to distinguish between the two subgroups. (i) Whereas members of both subgroups have been shown to possess Ser/Thr kinase activity, only members of subgroup 1 have been shown unequivocally to possess lipid kinase activity (see below). (ii) Members of each subgroup are more related in protein sequence across the kinase domain to fellow members of their respective group than they are to those of the other group. (iii) Members of subgroup 1 are of approx. 100 kDa in size, whereas members

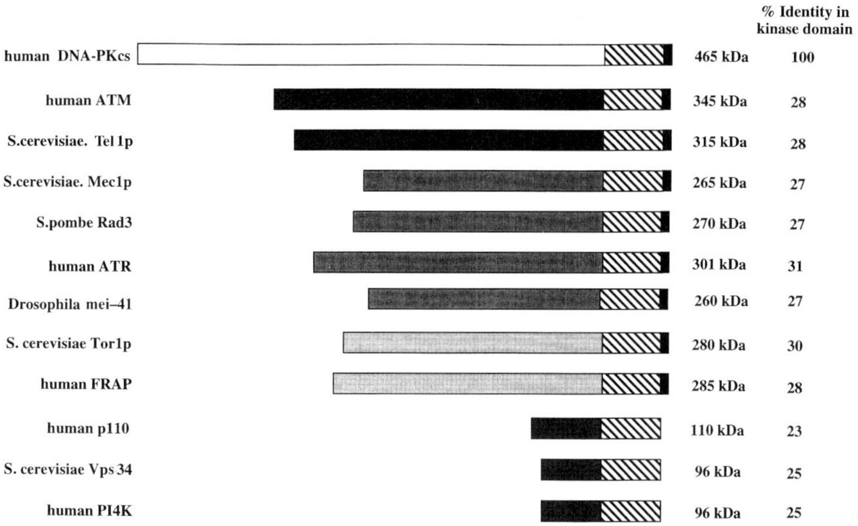

Fig. 2. Comparison of DNA-PK$_{cs}$ with other members of the PI 3-kinase family. Sequence comparisons were performed with the University of Wisconsin GCG package (Version 8) using the Smith and Waterman algorithm. The percentage identity over the last 400 amino acid residues (kinase domain, hatched region) of each protein in comparison with DNA-PK$_{cs}$ is shown on the right-hand side. This family can be further divided into two subfamilies. (i) The top nine proteins are all large (>260 kDa) and contain a C-terminal kinase domain extension (black rectangle). None of these proteins has been demonstrated unequivocally to exhibit lipid kinase activity (see the text for more details). (ii) The bottom three proteins are all ⩽110 kDa in size and do not contain a C-terminal extension. All members of this set of proteins tested have been demonstrated to exhibit lipid kinase function. Abbreviation: PI4K, PI 4-kinase.

of subgroup 2 are all >250 kDa in size. (iv) Members of subgroup 2 have a highly conserved C-terminal extension of 35 amino acid residues, while subgroup 1 does not contain this distal C-terminal region. (v) All members of subgroup 2 are involved in DNA repair, DNA-damage-sensing mechanisms and/or controlling cell-cycle progression.

The DNA-PK$_{cs}$ subgroup of the PI 3-kinase protein family

As described above, proteins containing the PI-3 kinase domain can be subdivided into two subgroups. The subgroup to which DNA-PK belongs has been the focus of considerable research in the last few years, with members of this family being shown to be involved in a variety of forms of genome surveillance and/or cell-cycle regulation. The yeast Tor proteins (Tor1p and Tor2p) and their mammalian homologue FRAP/mTOR were the first members of subgroup 2 to be cloned through studies on the mechanism of action of the macrolide antibiotic rapamycin [41–46]. When bound to the intracellular receptor FKBP12, rapamycin has the ability to inhibit the G1-to-S cell-cycle transition in both yeast and human cell types (reviewed in [47]). As a result of this inhibition of cell-cycle progression, rapamycin has anti-mitogenic and immunosuppressant functions, and this has led to its effective usage at a clinical level. Yeast strains that were generated to be resistant to rapamycin led to the identification of the two highly similar proteins, Tor1p and Tor2p (also termed Drr1p and Drr2p), as being the targets of the FKBP12–rapamycin complex. To isolate the mammalian target of the FKBP12–rapamycin complex, biochemical approaches were undertaken which resulted in the isolation of FRAP (i.e. **F**KBP12–**r**apamycin-**a**ssociated **p**rotein) [43–46]. That the kinase domain of these proteins is important for their function was demonstrated by mutagenisis of key residues in this region and studying the mutated proteins *in vivo* in yeast and mammalian cells.

Insight into how these PI-3-kinase-like proteins function *in vivo* has recently been obtained by the identification of an *in vivo* substrate for mTOR/FRAP [48]. Based on this result and other data, it appears that, by phosphorylating the translational repressor protein PHAS-I (phosphorylated heat- and acid-stable protein), mTOR/FRAP acts as the terminal kinase in a mitogenic signalling pathway. When phosphorylated by mTOR/FRAP, PHAS-I is abrogated in its association with the eukaryotic translation initiation factor eIF-4E. This dissociation of the PHAS-I–eIF-4E complex is thought to result in the stimulation of eIF-4E-dependent protein sythesis, which in turn leads to enhancement of the efficient translation of specific mRNAs that are crucial for efficient progression into S-phase. Therefore, through inhibiting this pathway, rapamycin prevents entry into S-phase.

Another member of subgroup 2 that has been the focus of considerable attention in recent years is the ATM protein, mutations in which lead to the human autosomal recessive disorder A-T. This disorder is present at an incidence of around 1 in 100 000 in the Western population and is characterized by a number of debilitating symptoms, including progressive cerebellar degeneration, oculo-cutaneous telangiectasia, growth retardation, immune deficiency, increased radiosensitivity and aspects of premature aging (reviewed in [49–51]). A-T patients also exhibit an approx. 100-fold increased incidence of cancer, with patients being particularly predisposed to malignancies of lymphoid origin. Furthermore, A-T heterozygotes, which are estimated to represent approx. 0.5% of the population, are reported to exhibit a higher incidence of breast cancer (reviewed in [49]). At the cellular level, A-T is characterized by a high degree of chromosomal instability, radio-resistant DNA synthesis and hypersensitivity to ionizing radiation and radio-mimetic drugs; this is manifested, at least in part, in defective cell-cycle checkpoint controls in response to these DNA-damaging agents. For example, A-T cells are deficient in effecting the G1–S and G2–M cell-cycle checkpoints following treatment with ionizing radiation. One defining characteristic of A-T cells is that they exhibit radio-resistant DNA synthesis, indicating a defect in the radiation-induced S-phase cell-cycle checkpoint. A-T patients, or A-T knockout mice, also exhibit completely deficient or delayed p53 activation in response to ionizing radiation [52–55]. Thus it is thought that the A-T gene product acts upstream of p53 in the signalling pathway activated in response to ionizing radiation.

After mapping of the locus mutated in A-T patients to human chromosome 11q22-23 [56], the gene called *ATM* (A-T mutated) was identified and the cDNA corresponding to this locus was cloned [29]. Sequence analyses revealed that the *ATM* gene encodes a polypeptide of approx. 350 kDa that is a member of the DNA-PK_{cs} subgroup of the PI 3-kinase family of proteins [29]. Although cloned over 2 years ago, very little is yet known about the biochemistry of ATM or of the molecular details of its mechanism of action. However, work from our laboratory and others has revealed that the ATM protein appears to be predominantly nuclear, and its level of expression and cellular location appear not to be modulated in response to ionizing radiation [57–60]. These observations therefore provide indirect evidence for models in which ATM acts as part of an apparatus that monitors genomic integrity. Biochemical analysis of the ATM protein is under way in our laboratory, and we have recently purified this protein to homogeneity and have also found that it binds a variety of different DNA substrates (results not shown). We hope that this biochemical approach will help in our understanding of how ATM functions in the nucleus. Purification of ATM, we hope, will also lead us to identify associated proteins. In light of the DNA-PK paradigm, it is tempting to speculate that such proteins may play a role in directing ATM to aberrations in the genome and thus trigger its kinase activity on appropriate DNA structures.

When the ATM sequence is compared with those of the other members of subgroup 2, it is seen to share greatest identity with the *S. cerevisiae TEL1* gene product [33,34]. *TEL1* was identified originally through genetic screens in an attempt to isolate yeast strains that possess shortened telomeres. Not only is this phenotype found for *tel1* mutants, but these yeasts also have enhanced rates of chromosomal loss and exhibit increased mitotic recombination. However, unlike A-T cells, *tel1* yeast strains are not hypersensitive to DNA-damaging agents such as ionizing radiation and do not appear to have aberrent cell-cycle checkpoints. Nevertheless, Tel1p overexpression complements a *mec1* mutant, indicating that Tel1p and Mec1p are likely to have overlapping and partially redundant functions in maintaining genomic stability. Furthermore, in keeping with telomeric defects of yeast *tel1* mutants, it has been reported that A-T cells possess shortened telomeres [61,62].

Also showing greater identity with ATM than with DNA-PK$_{cs}$ and the Tor/FRAP proteins are *S. cerevisiae* Mec1p (also called Esr1p) and *Schiz. pombe* Rad3. Mec1p and Rad3 were isolated via genetic screens to identify genes that play an important role in DNA repair and DNA-damage-induced cell-cycle checkpoint controls [35–38]. These two proteins display considerable identity, even outside the kinase domain, and are believed to be homologues of one other, along with the human ATR protein [39] and *mei-41* [40]. Significantly, *rad3* and *mec1* mutant strains are hypersensitive to UV light, ionizing radiation and a range of DNA-damaging chemical agents, such as methyl methanesulphonate. One apparent reason for the sensitivity of *rad3* and *mec1* strains to these genomic insults is that they are unable to mediate the G1–S, S and G2–M checkpoints after mutagen exposure. Interestingly, *mec1* and *rad3* mutant yeasts also lack the checkpoint mechanism which ensures that entry into mitosis occurs after the completion of DNA replication. Hence *rad3* and *mec1* mutant strains will attempt to enter into mitosis in the presence of stalled replication forks. Similarly to *mec1* or *rad3* mutant yeast, *mei-41 Drosophila* mutants show increased genomic instability and hypersensitivity to DNA-damaging agents, and are unable to delay entry into mitosis after exposure to genomic insults [40]. *mei-41* was originally identified in a screen for mutants defective in meiotic recombination, a phenotype that is shared by *rad3* and *mec1* mutant yeasts.

DNA-PK is a protein kinase but appears to lack lipid kinase activity

The fact that DNA-PK$_{cs}$, and the other proteins described above, display strong sequence similarity to the catalytic region of lipid kinases (Fig. 2) has led us to question whether DNA-PK possesses lipid kinase activity. Members of the lipid kinase family phosphorylate phosphoinositides on the inositol ring, after being activated by a range of extracellular stimuli such as the activation of growth factor receptors. This leads to the rapid production of small, potent signalling lipids that act as intracellular second messengers, and eventually to an appropriate cellular response to the original stimulus. If DNA-PK is able to phosphorylate lipids, then a clear role for this protein in a nuclear phosphatidylinositol signalling pathway could be envisaged. Initial studies did reveal that a weak PI 4-kinase activity was present in preparations of DNA-PK purified from HeLa cell nuclear extracts [27]. However, this activity did not purify further with DNA-PK-associated protein kinase activity. As can be seen in Fig. 3, a highly purified preparation of DNA-PK has lost this PI 4-kinase activity, but retains DNA-activatable protein kinase activity. Further purification of DNA-PK into its component parts of Ku and DNA-PK_{cs} revealed that, upon reconstitution, protein kinase activity was established, but not lipid kinase activity (Fig. 4). It should be noted that the DNA-PK_{cs} used here does not contain any protein kinase potential in the absence of the DNA-targeting subunit Ku (results not shown). In line with the above, substrate inhibition experiments suggested that the PI 4-kinase activity detected in partially pure DNA-PK preparations was a contaminant and that the DNA-PK holoenzyme is not a *bona fide* lipid kinase (results not shown).

As well as PI, other potential substrates for DNA-PK have been tested, including diacylglycerol, PI 4-phosphate and PI 4,5-bisphosphate. However, in all cases we have found that these are not phosphorylated by our highly purified DNA-PK preparations. Since it has been demonstrated that the Mn^{2+} ion is required for efficient phosphorylation by the p110 subunit of human PI 3-kinase, we also performed all assays in both the

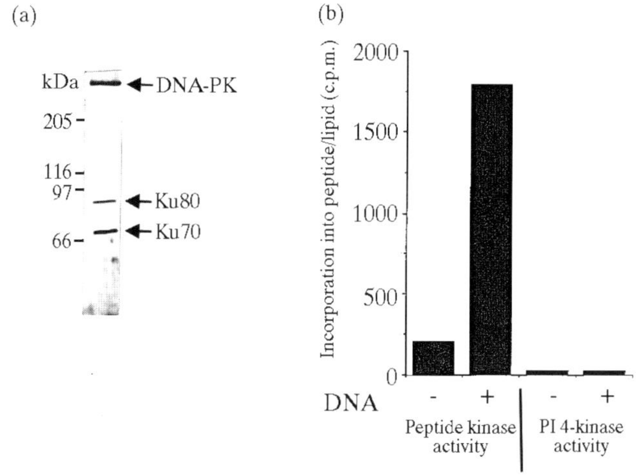

Fig. 3. Highly purified DNA-PK does not exhibit PI 4-kinase activity. (a) DNA-PK holoenzyme was purified from HeLa nuclear extract by Q-Sepharose, S-Sepharose, heparin–agarose and double-stranded-DNA–cellulose chromatography. Purified protein was visualized by silver staining of a 7% denaturing polyacrylamide gel. (b) Purified DNA-PK was utilized in peptide kinase and lipid kinase assays as previously described [27]. Assays were performed in either the absence (−) or the presence (+) of sonicated calf thymus DNA, as indicated.

presence and the absence of either Mg^{2+} or Mn^{2+}. Finally, to date, no intrinsic lipid kinase activity has been shown unequivocally to be associated with any of the other members of subgroup 2 of the proteins shown in Fig. 2. Of course, the possibility cannot be ruled out that, as yet, we have not used the correct lipid substrate for DNA-PK. Nevertheless, taken together, the available data suggest strongly that DNA-PK, and most likely other members of the DNA-PK_{cs} subgroup of the PI 3-kinase family, are protein kinases but not lipid kinases.

DNA-PK is inhibited by the PI 3-kinase inhibitors wortmannin and LY294002

Wortmannin is a low-molecular-mass hydrophobic compound, with a sterol-like structure (Fig. 5), which was isolated as a fungal metabolite (reviewed in [63]). Wortmannin was first identified as an inhibitor of myosin light chain kinase, with an IC_{50} of about 2.5 μM [64]. However, in the past few years it has become clear that wortmannin is a much more potent inhibitor of mammalian PI 3-kinase, and consequently it has become an important tool in the study of PI 3-kinase function [63]. Wortmannin inhibits mammalian PI 3-kinases with an IC_{50} of approx. 2.5 nM, with slight variations depending on the system being studied. The

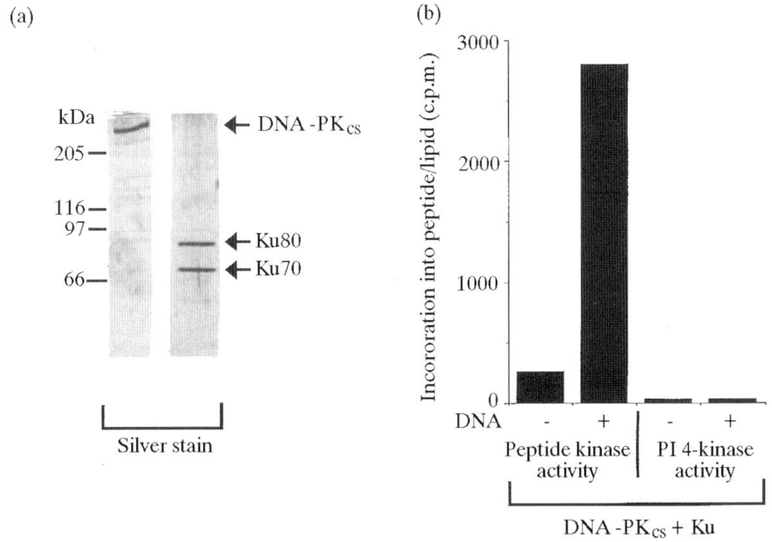

Fig. 4. Reconstituted DNA-PK does not exhibit PI 4-kinase activity. (a) Purified DNA-PK holoenzyme (as in Fig. 2) was separated into DNA-PK_{cs} and Ku via chromatography on phenyl-Sepharose. (b) Reconstitution of DNA-PK_{cs} and Ku results in kinase activity towards a peptide substrate, but not towards PI. Assays were performed in either the absence (−) or the presence (+) of sonicated calf thymus DNA, as indicated.

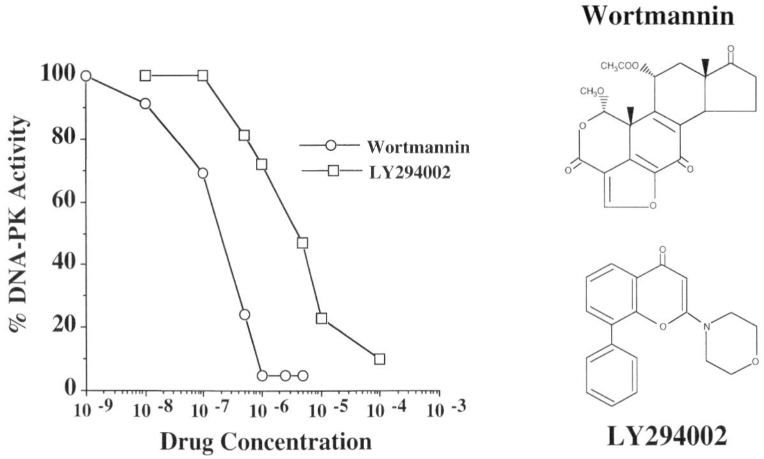

Fig. 5. DNA-PK activity is inhibited by the PI 3-kinase inhibitors wortmannin and LY294002. DNA-PK peptide phosphorylation assays were performed in the presence of various concentrations of wortmannin or LY294002, as indicated. From the dose–response curves generated, IC_{50} values were derived: wortmannin, 200 nM; LY294002, 2.5 μM.

binding of wortmannin to the p110 subunit of PI 3-kinase is covalent in nature, thus displaying non-competitive inhibition [65]. The binding of wortmannin to the active site of the p110 subunit of PI 3-kinase has recently been demonstrated [65]. More recently, another inhibitor of mammalian PI 3-kinases, LY294002, has been identified and has been used in studies of the mammalian PI 3-kinases [66]. This synthetic compound appears to have an IC_{50} of 1.4 μM against the PI 3-kinases studied so far, with the inhibition appearing to be of a competitive nature [66].

Because of the sequence similarities between the catalytic domains of DNA-PK_{cs} and p110, we tested whether DNA-PK activity could be inhibited by wortmannin and/or LY294002. Figure 5 shows dose–response curves of DNA-PK activity in the presence of increasing concentrations of each compound. These data reveal that wortmannin has an IC_{50} of around 200 nM, while LY294002 has an IC_{50} of around 2.5 μM. Although the IC_{50} for wortmannin appears to be two orders of magnitude higher than that for classical PI 3-kinases, the IC_{50} for LY294002 appears very similar to those identified for the other PI 3-kinases. This serves to highlight the point that these compounds should no longer be considered as 'specific inhibitors of PI 3-kinase'. Since DNA-PK and the p110 subunit of PI 3-kinase are members of an ever-growing family of proteins, it will be important to identify the effects of these compounds on the other members of this family. However, what this information does provide is the starting point for drug discovery with the aim of identifying compounds that may be more specific for one family member than another, and which may modulate the physiological process that the respective kinase is involved in. Intriguingly, it has been found that, at high doses (10–50 μM), wortmannin can act as a sensitizer of cells to ionizing radiation [67,68]. Furthermore, the data from one of these studies suggests that wortmannin is a potent inhibitor of the repair of DNA double-strand breaks, but not single-strand breaks, in Chinese hamster ovary cells [68]. However, it is unknown which kinase is being inhibited by wortmannin in these studies, but it is tempting to speculate that DNA-PK or ATM may be the target. To back this idea up, a recent paper has suggested that DNA end-joining activity in *Xenopus* extracts is inhibited by wortmannin and may be mediated via DNA-PK [69].

Concluding remarks

Studies on the biochemistry of DNA-PK have provided us with valuable insights into the role of this molecule in the processes of DNA repair and recombination. Elucidation of the mechanism of action of this kinase may also provide us with informative insights into how other members of subgroup 2 of the PI 3-kinase family of proteins function at the molecular level. Furthermore, identification of specific inhibitors of DNA-PK and ATM may lead to effective reagents that can sensitize tumour cells to the effects of ionizing radiation and/or radio-mimetic drugs. In the light of their clear importance, it seems obvious that DNA-PK and its relatives are going to be the subject of much intensive future research.

References

1. Walker, A.I., Hunt, T., Jackson, R.L. and Anderson, C.W. (1985) EMBO J. **4**, 139–145
2. Jackson, S.P., MacDonald, J.J., Lees-Miller, S. and Tjian, R. (1990) Cell **63**, 155–165
3. Lees-Miller, S.P., Chen, Y.-R. and Anderson, C.W. (1990) Mol. Cell. Biol. **10**, 6472–6481
4. Carter, T., Vancurova, I., Sun, I., Lou, W. and DeLeon, S. (1990) Mol. Cell. Biol. **10**, 6460–6471
5. Lees-Miller, S.P., Sakaguchi, K., Ullrich, S.J., Appella, E. and Anderson, C.W. (1992) Mol. Cell. Biol. **12**, 5041–5049
6. Anderson, C.W. and Lees-Miller, S.P. (1992) Crit. Rev. Eukaryotc Gene Expression **2**, 283–314
7. Gottlieb, T.M. and Jackson, S.P. (1993) Cell **72**, 131–142
8. Dvir, A., Peterson, S.R., Knuth, M.W., Lu, H. and Dynan, W.S. (1992) Proc. Natl. Acad. Sci. U.S.A. **89**, 11920–11924
9. Morozov, V.E., Falzon, M., Anderson, C.W. and Kuff, E.L. (1994) J. Biol. Chem. **269**, 16684–16688
10. Bannister, A.J., Gottlieb, T.M., Kouzarides, T. and Jackson, S.P. (1993) Nucleic Acids Res. **21**, 1289–1295
11. Anderson, C.W. (1993) Trends Biochem. Sci. **18**, 433–437
12. Jackson, S.P. (1996) Cancer Surv, **28**, 261–279
13. Jackson, S.P. (1997) Int. J. Biochem. Cell. Biol. **7**, 935–938

14. Kuhn, A., Gottlieb, T.M., Jackson, S.P. and Grummt, I. (1995) Genes Dev. **9**, 193–203
15. Labhart, P. (1995) Proc. Natl. Acad. Sci. U.S.A. **92**, 2934–2938
16. Jeggo, P.A., Taccioli, G.E. and Jackson, S.P. (1995) BioEssays **17**, 949–957
17. Jackson, S.P. and Jeggo, P.A. (1995) Trends Biochem. Sci. **20**, 412–415
18. Lieber, M.R., Grawunder, U., Wu, X. and Yaneva, M. (1997) Curr. Opin. Genet. Dev. **7**, 99–104
19. Zhu, C., Bogue, M.A., Lim, D.-S., Hasty, P. and Roth, D.B. (1996) Cell **86**, 379–389
20. Wu, X. and Lieber, M.R. (1996) Mol. Cell. Biol. **16**, 5186–5193
21. Boulton, S.J. and Jackson, S.P. (1996) EMBO J. **15**, 5093–5103
22. Boulton, S.J. and Jackson, S.P. (1996) Nucleic Acids Res. **24**, 4639–4648
23. Milne, G.T., Jin, S., Shannon, K.B. and Weaver, D.T. (1996) Mol. Cell. Biol. **16**, 4189–4198
24. Siede, W., Friedl, A.A., Dianova, I., Eckardt-Schpp, F. and Friedberg, E.C. (1996) Genetics **142**, 91–102
25. Mages, G.J., Feldmann, H.M. and Winnacker, E.-L. (1996) J. Biol. Chem **271**, 7910–7915
26. Porter, S.E., Greenwell, P.W., Ritchie, K.B. and Petes, T.D. (1996) Nucleic Acids Res. **24**, 582–585
27. Hartley, K.O., Gell, D., Smith, G.C.M., Zhang, H., Divecha, N., Connelly, M.A., Admon, A., Lees-Miller, S.P., Anderson, C.W. and Jackson, S.P. (1995) Cell **82**, 849–856
28. Zakian, V.A. (1995) Cell **82**, 685–687
29. Savitsky, K., Bar-Shira, A., Gilad, S., Rotman, G., Ziv, Y., Vanagaite, L., Tagle, D.A., Smith, S., Uziel, T., Sfez, S., et al. (1995) Science **268**, 1749–1753
30. Kapeller, R. and Cantley, L.C. (1994) BioEssays **16**, 565–576
31. Carpenter, C.L. and Cantley, L.C. (1996) Curr. Opin. Cell Biol. **8**, 153–158
32. Keith, C.T. and Schreiber, S.L. (1995) Science **270**, 50–51
33. Greenwell, P.W., Kronmal, S.L., Porter, S.E., Gassenhuber, J., Obermaier, B. and Petes, T.D. (1995) Cell **82**, 823–829
34. Morrow, D.M., Tagle, D.A., Shiloh, Y., Collins, F.S. and Heiter, P. (1995) Cell **82**, 831–840
35. Kato, R. and Ogawa, H. (1994) Nucleic Acids Res. **22**, 3104–3112
36. Weinert, T.A., Kiser, G.L. and Hartwell, L.H. (1994) Genes Dev. **8**, 652–665
37. Jimenez, G., Yucel, J., Rowley, R. and Subramani, S. (1992) Proc. Natl. Acad. Sci. U.S.A. **89**, 4952–4956
38. Seaton, B.L., Yucel, J., Sunnerhagen, P. and Subramani, S. (1992) Gene **119**, 83–89
39. Cimprich, K., Shin, T.B., Keith, C.T. and Schreiber, S.L. (1996) Proc. Natl. Acad. Sci. U.S.A. **93**, 2850–2855
40. Hari, K.L., Santerre, A., Sekelsky, J.J., McKim, K.S., Boyd, J.B. and Hawley, R.S. (1995) Cell **82**, 815–821
41. Heitman, J., Movva, N.R. and Hall, M.N. (1991) Science **253**, 905–909
42. Kunz, J., Henriquez, R., Schneider, U., Deuter-Reinhard, M., Movva, N.R. and Hall, M.N. (1993) Cell **73**, 585–596
43. Brown, E.J., Albers, M.W., Shin, T.B., Ichikawa, K., Keith, C.T., Lane, W.S. and Schreiber, S.L. (1994) Nature (London) **369**, 756–758
44. Chiu, M.I., Katz, H. and Berlin, V. (1994) Proc. Natl. Acad. Sci. U.S.A. **91**, 12574–12578
45. Helliwell, S.B., Wagner, P., Kunz, J., Deuter-Reinhard, M., Henriquez, R. and Hall, M.N. (1994) Mol. Biol. Cell **5**, 105–118

46. Sabatini, D.M., Erdjument-Bromage, H., Liu, M., Tempst, P. and Snyder, S.H. (1994) Cell **78**, 35–43
47. Brown, E.J. and Schreiber, S.L. (1996) Cell **86**, 517–520
48. Brunn, G.J., Hudson, C.C., Sekulic, A., Williams, J.M., Hosoi, H., Houghton, P.J., Lawrence, J.C. and Abraham, R.T. (1997) Science **277**, 99–101
49. Meyn, M.S. (1995) Cancer Res. **55**, 5991–6001
50. Shiloh, Y. (1995) Eur. J. Hum. Genet. **3**, 116–138
51. Jackson, S.P. (1995) Curr. Biol. **5**, 1210–1212
52. Kastan, M.B., Zhan, Q., El-Deiry, W.S., Carrier, F., Jacks, T., Walsh, W.V., Plunkett, B.S., Vogelstein, B. and Fornace, A.J.J. (1992) Cell **71**, 587–597
53. Lu, X. and Lane, D.P. (1993) Cell **75**, 765–778
54. Khanna, K. and Lavin, M.F. (1993) Oncogene **8**, 3307–3312
55. Xu, Y. and Baltimore, D. (1996) Genes Dev. **10**, 2401–2410
56. Gatti, R.A., Berkel, I., Boder, E., Braedt, G., Charmley, P., Concanon, P., Ersoy, F., Foroud, T., Jasper, N.G., Lange, K., et al. (1988) Nature (London) **336**, 577–580
57. Chen, G. and Lee, E.Y.-H.P. (1996) J. Biol. Chem. **271**, 33693–33697
58. Lakin, N.D., Weber, P., Stankovic, T., Rottinghaus, S.T., Taylor, A.M. and Jackson, S.P. (1996) Oncogene **13**, 2707–2716
59. Brown, K.D., Ziv, Y., Sadanandan, S.N., Chessa, L., Collins, F.S., Shiloh, Y. and Tagle, D.A. (1997) Proc. Natl. Acad. Sci. U.S.A. **94**, 1840–1845
60. Watters, D., Khanna, K.K., Beamish, H., Birrell, G., Spring, K., Kedar, P., Gatei, M., Stenzel, D., Hobson, K., Kozlov, S., et al. (1997) Oncogene **14**, 1911–1921
61. Pandita, T.K., Pathak, S. and Geard, C.R. (1995) Cytogenet. Cell. Genet. **71**, 86–93
62. Metcalfe, J.A., Parkhill, J., Campbell, L., Stacey, M., Biggs, P., Byrd, P.J. and Taylor, A.M.R. (1996) Nature Genet. **13**, 350–353
63. Ui, M., Okada, T., Hazeki, K. and Hazeki, O. (1995) Trends Biochem. Sci. **20**, 303–307
64. Nakanishi, S., Kakita, S., Takahashi, I., Kawahara, K., Tsukudu, E., Sano, T., Yamada, K., Yoshida, M., Kase, H., Matsuda, Y., et al. (1992) J. Biol. Chem. **267**, 2157–2163
65. Wymann, M.P., Bulgarelli-Leva, G., Zvelebil, M.J., Pirola, L., Vanhaesebroeck, B., Waterfield, M.D. and Panayotou, G. (1996) Mol. Cell. Biol. **16**, 1722–1733
66. Vlahos, C.J., Matter, W.F., Hui, K.Y. and Brown, R.F. (1994) J. Biol. Chem. **269**, 5241–5248
67. Price, B.D. and Youmell, M.B. (1996) Cancer Res. **36**, 246–250
68. Boulton, S., Kyle, S., Yalcintepe, L. and Durkacz, B.W. (1996) Carcinogenesis **17**, 2285–2290
69. Gu, X., Bennet, R.A.O. and Povirk, L.F. (1996) J. Biol. Chem. **271**, 19660–19663

Biochem. Soc. Symp. **64**, 105–118
Printed in Great Britain

8

Stress-induced activation of the heat-shock response: cell and molecular biology of heat-shock factors

Jose J. Cotto* and Richard I. Morimoto†

Department of Biochemistry, Molecular Biology and Cell Biology, Rice Institute for Biomedical Research, Northwestern University, 2153 North Campus Drive, Evanston, IL 60208, U.S.A.

Abstract

Exposure of cells to environmental and physiological stress leads to an imbalance in protein metabolism, which challenges the cell to respond rapidly and precisely to the deleterious effects of stress on protein homoeostasis. The heat-shock response, through activation of heat-shock transcription factors (HSFs) and the elevated expression of heat-shock proteins and molecular chaperones, protects the cell against the accumulation of non-native proteins. Activation of HSF1 involves a complex multistep pathway in which the inert monomer oligomerizes to a DNA-binding, transcriptionally active, trimer which relocalizes within the the nucleus to form stress-induced HSF1 granules. Attenuation of the heat-shock response involves molecular chaperones which repress the HSF1 transactivation domain and HSF-binding protein 1 (HSBP1), which interacts with the HSF1 oligomerization domain of HSF1 to negatively regulate its activity, thus insuring that the expression of chaperones is precisely determined.

Introduction

Every cell responds to environmental, chemical and physiological stress through a rapid and preferential increase in expression of a highly conserved group of proteins known as the heat-shock proteins (Hsps). The cellular response to stress was discovered by Ritossa [1], who

*Present address: Department of Chemical Engineering, University of Texas at Austin, Austin, TX 78712, U.S.A.
†To whom correspondence should be addressed.

observed an induction of specific chromosomal puffs in the polytene chromosomes of *Drosophila melanogaster* upon exposure to elevated temperatures and chemical treatment. Concomitantly these treatments also caused a reduction in the number of pre-existing puffs, suggesting that the heat-shock response results in the reprogramming of the overall pattern of gene transcription [2–4].

Subsequent studies elucidated the nature of the induced RNAs and proteins at a molecular level and led to the isolation and characterization of the Hsp genes [5,6]. Hsps have been categorized into several families according to their molecular mass, and each member has an essential role as a molecular chaperone in the processes of protein synthesis, translocation, folding or degradation [7–12].

The exact mechanism by which the cell recognizes stress in order to activate the heat-shock response remains unclear. Studies carried out in diverse plants and animals have revealed that the expression of Hsps is inducibly regulated by a number of seemingly unrelated conditions that can be categorized into three major classes: environmental stress, physiological stress and non-stressful conditions (Fig. 1). A characteristic shared

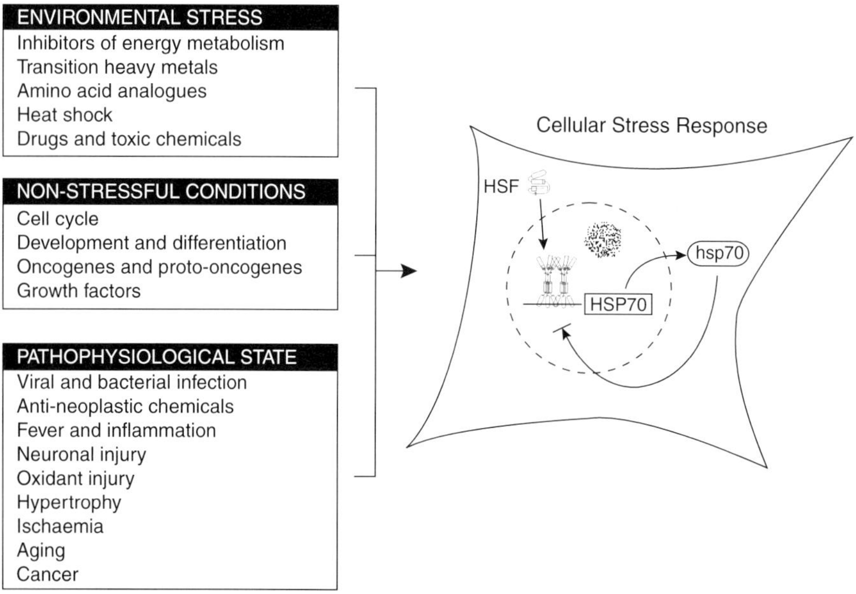

Fig. 1. Conditions that induce heat-shock gene expression. Three classes of cellular conditions are indicated: (1) environmental stress; (2) non-stressful conditions; and (3) pathophysiological state. Representative examples of inducers of heat-shock gene expression in each class are listed. The diagram corresponds to the stress-dependent activation of HSF which occurs in response to environmental and pathophysiological stress and results in elevated expression of Hsp70. Reproduced from Morimoto, R.I., et al. (1997) Essays Biochem. **32**, 17–29 (Portland Press, London), with the permission of the Biochemical Society.

by some but not all of these conditions is their ability to induce protein denaturation and/or aggregation. Hence the heat-shock response has been considered as a safeguard mechanism that ensures survival of the cell in the presence of otherwise lethal forms of stress that induce protein damage [13–15].

In eukaryotic cells, heat-shock genes are under the control of a family of conserved DNA-binding proteins collectively referred to as the heat-shock transcription factors (HSFs) [16,17]. The transcriptional activation of heat-shock genes by HSFs is an excellent model for the study of inducible gene expression, and its complexity is exemplified by the diverse array of conditions that induce the response. Thus an understanding of the heat-shock response requires an appreciation of the myriad conditions that lead to the elevated expression of Hsps. This chapter discusses the regulation of the HSF multi-gene family and the role of these transcriptional activators in the inducible expression of genes encoding Hsps and molecular chaperones.

A multi-gene family of HSFs

An unexpected finding during the cloning of the HSF genes from plants and vertebrates was the identification of an HSF multi-gene family. At least three HSFs have been isolated from the human, mouse, chicken and tomato genomes and an additional factor, HSF4, has been identified in humans [18–23]. Comparison of the structures and sequences of the vertebrate HSFs (Fig. 2) reveals that, within a single species, the HSFs are approx. 40% related in amino acid sequence (i.e. mouse HSF1/HSF2 or chicken HSF1/HSF2/HSF3), which is primarily due to the extensive identity of the DNA-binding and oligomerization domains. Interspecies comparisons (i.e. between human, mouse and chicken HSF1) indicates approx. 92% conservation in sequence (Table 1). Comparison of the three chicken HSFs with other cloned HSFs suggests that there was a common ancestor from which they diverged [18]. By comparison, the tomato HSFs (HSF8, HSF24 and HSF30), although similar in structure, were isolated independently through binding site screening; furthermore, the plant HSFs do not share the same family relationships as have been established for the vertebrate HSFs [22].

What is the role of multiple HSFs? One possibility is that larger, more complex organisms may utilize multiple HSFs to respond to an increasingly diverse array of developmental and environmental cues. Another consideration is that the gene encoding *Saccharomyces cerevisiae* HSF, which is an essential gene, has at least two transcriptional activation domains that respond to sustained or transient heat shock [24–26]. Perhaps the duplication of HSFs in the larger eukaryotes was a response to evolutionary events that required the co-regulation of the heat-shock genes in response to distinct signals. An answer to this question was provided by the demonstration that HSF1 corresponds to the rapidly activated stress-responsive factor, whereas the co-expressed HSF2 is activated in response to distinct developmental cues [27,28]. During haemin treatment of K562 erythroleukaemia cells, which leads to non-terminal erythroid differentiation, HSF2 is converted from an inert dimer into an active trimer [29,30]. An important distinction, however, is that HSF1 undergoes rapid (within seconds) activation in response to heat and attenuates rapidly, whereas haemin-induced HSF2 is activated over a period of time, requiring 16–24 h and remaining activated for up to 72 h. Simultaneous activation of HSF1 and HSF2 in K562 cells by exposure of haemin-treated cells to heat shock led to a synergistic transcriptional activation of the *Hsp70* gene, suggesting that HSF1 and HSF2 can co-exist in their respective activated states. Other studies on mouse tissues revealed

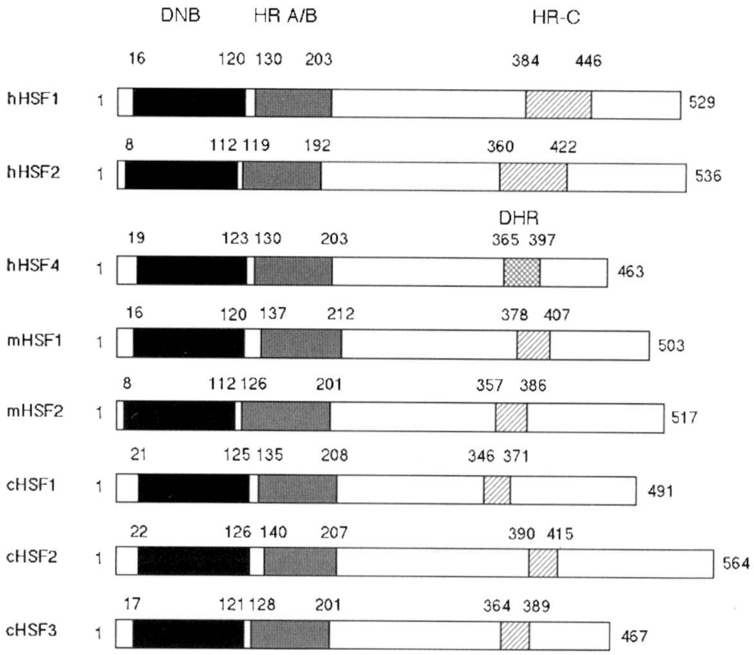

Fig. 2. HSF gene family and structural domains. General features of human (h) HSF1, HSF2 and HSF4, mouse (m) HSF1 and HSF2, and chicken (c) HSF1, HSF2 and HSF3 are presented schematically. The DNA-binding domain (DNB) of approx. 100 amino acids is localized to the N-terminus (black box). Adjacent is the conserved hydrophobic heptad repeats (HR-A/B) responsible for HSF trimerization (shaded box). The C-terminus contains an additional heptad repeat (HR-C) involved in negative regulation (cross-hatched box). hHSF4 does not have a region corresponding to HR-C. Instead, a variant of the C-terminus hydrophobic repeat is present, which is referred to as the downstream of heptad repeat (DHR) region. The relative location (in amino acids relative to the N-terminus) of each conserved region is indicated.

Table 1. Biochemical and cell biological characteristics of the HSF protein family. Analysis of HSF4 is based on transient or stably transfected cells. Abbreviation: cyto., cytoplasmic.

	HSF1		HSF2			HSF3		HSF4
Species	Human, mouse, chicken*		Human, mouse, chicken*			Chicken		Human
Expression	Constitutive		Constitutive			Constitutive		Tissue-specific (heart, skeletal muscle, brain)
Conditions	37°C	42°C	37°C	42°C	Haemin	37°C	45°C	37°C†
Molecular mass (kDa)								
Native	70	178	127	127	202			
Denatured	70	85	72	72	72	69	69	55
Subcellular localization	Cyto./nuclear	Nuclear	Cyto./nuclear	Cyto./nuclear	Nuclear	Cyto.	Nuclear	Constitutively nuclear
Oligomeric state	Monomer	Trimer	Dimer	Dimer	Trimer	Dimer	Trimer	Trimer
DNA binding	—	+	—	—	+	—	+	Constitutive DNA binding but lacks transcriptional activity
Biochemical modifications	Constitutive phosphorylation	Inducible phosphorylation	None	None	None	None	None	None

*Identity = 92% between species.
†DNA-binding activity is lost *in vitro* upon heat shock.

that HSF2 mRNA is developmentally regulated and accumulates to high levels in the testis at the spermatocyte and round spermatid stages of development [31,32].

Studies on chicken HSF3 revealed additional diversity in the pathways for activation of HSFs [18,33]. HSF3 is expressed ubiquitously; however, its activity is induced upon heat shock in chicken erythroblastic cells (HD6) and not in cells from other lineages. HSF3 has many characteristics similar to those of HSF1, such as sequence-specific binding to the heat-shock element, negative regulation and activation to a trimer; however, unlike HSF1, HSF3 is an inert dimer. As for HSF1, deletion of leucine zipper 4 resulted in constitutive DNA-binding activity, which suggests a role for zipper 4 in the negative regulation of HSF3 DNA-binding activity [18]. HSF3 has the activity of a positive activator, as demonstrated by co-transfection with a reporter gene. These data suggest that HSF1 and HSF3 could enhance the ability of the cell to tightly regulate the heat-shock response in a cell-type-specific manner [33]. Human HSF4 represents a novel member of the HSF family which appears to lack a trans-activation domain and has properties of a repressor of heat-shock gene expression [19].

Mechanisms of HSF regulation

Regulation of the oligomeric state

A prominent regulatory feature of HSFs is oligomerization to an active trimeric DNA-binding state. *Drosophila* and vertebrate HSFs are maintained in a latent monomeric form and converted into a DNA-binding competent trimer upon heat shock [34]. The biochemical events associated with oligomerization have been studied through the use of a variety of biochemical methods, including chemical cross-linking reagents and hydrodynamic studies. The latent HSF molecule in *D. melanogaster* is a very elongated monomer with a frictional ratio (f/f_0) of approx. 1.9 [35], whereas studies with human HSF1 and HSF2 have demonstrated that the f/f_0 for monomeric HSF1 is approx. 1.7 and that for HSF2 is also approx. 1.7 [29]. Upon conversion into the trimeric form during heat shock, the frictional ratio for *Drosophila* HSF increases (f/f_0 2.6), suggesting an unfolding event in conjunction with the trimerization, whereas the vertebrate HSF1 is essentially unaffected. The lack of a significant change in the axial ratio of the vertebrate HSF may be due to its smaller size; human HSF1 and HSF2 have molecular masses of 54 and 58 kDa respectively, whereas that of *Drosophila* HSF is 110 kDa.

Investigation of the physical parameters of HSF2 from K562 cells revealed, surprisingly, that it is a dimer in the latent form and is converted upon haemin treatment into an active DNA-binding trimer [29]. The difference in the oligomeric states of latent HSF1 and HSF2 corroborates the observed differences in activation in response to different signalling pathways.

A central question has been to understand how HSF is maintained in the latent state. Studies on *Drosophila* HSF have suggested a role for intramolecular interactions between heptad repeat C (HR-C), at the C-terminus of the protein, and the heptad repeats of the oligomerization domain to form the inert monomer [36]. In these studies, mutation of HR-C resulted in constitutive DNA-binding activity. These results were corroborated by an analysis of HSF3 which demonstrated that deletion of HR-C also resulted in constitutive DNA-binding activity [18]. Taken together, these studies suggest that intramolecular interactions between the N- and C-terminal leucine zippers regulate HSF1 activity, and are consistent with the unfolding events observed during heat-shock activation (Fig. 3).

Transcription activation domains of HSF1

Trimerization of HSF is essential to its function and provides the high-affinity DNA binding required for interaction with the heat-shock element. When bound to the DNA through the N-terminal DNA-binding domain, HSF is likely to affect transcriptional activation via sequences in the C-terminus. The activation domains of the yeast (*S. cerevisiae* and *Kluyveromyces lactis*) HSFs and the tomato HSFs have been dissected, revealing differences in the nature of the respective activation domains [24,37]. Yeast HSF has multiple transcriptional activation domains that mediate the response to sustained and transient heat shock. The N-terminal activator of *S. cerevisiae* is responsible for activation of transcription during sustained stress [25,26], whereas the C-terminal activator responds to transient stress. The *S. cerevisiae* C-terminal activation domain is approx. 180 amino acids long and appears to be bipartite, with one domain (amino acids 595–713) responsible for activation and the adjacent sequences (amino acids 713–780) having a supporting function for the activator. In contrast, the corresponding region in *K. lactis* HSF is a 32-amino-acid sequence (residues 592–623) with some α-helical character and acidic quality. The tomato HSFs apparently have a third type of activation motif located in the C-terminus which is composed of acidic sequences that contain a central tryptophan residue.

The activation domains of mammalian HSFs (human and mouse) were characterized through the use of chimaeric factors in which the heterologous GAL4 DNA-binding domain was fused to various segments of HSF1 to assess the activity of respective HSF sequences. Such studies have identified two activation domains, referred to as AD1 and AD2, both of which are localized at the extreme C-terminus of HSF1 [38–40]. Fusion of the GAL4 DNA-binding domain to residues 124–503 of HSF1 results in a chimaeric factor that binds DNA yet lacks any transcriptional activity. In order to detect transcriptional activity from GAL4–HSF1, it was necessary to delete an array of leucine zippers (zippers 1–3) positioned between residues 124 and 227 of HSF1. The minimum negative regulatory domain was shown to require residues between amino acids 188 and 227. The minimal region for transcriptional activation was mapped to

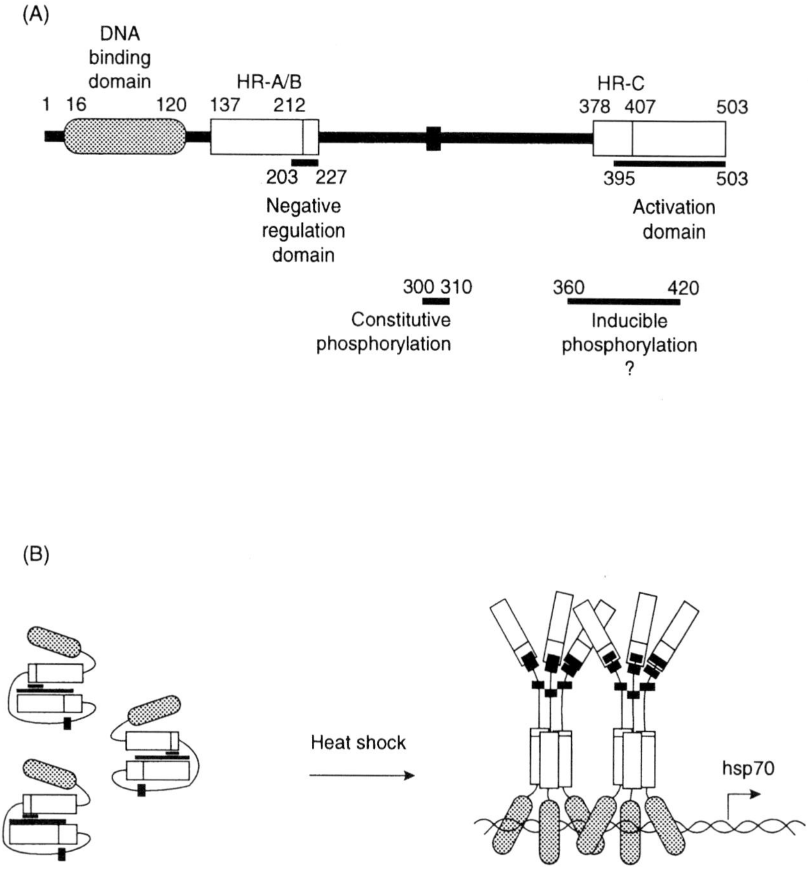

Fig. 3. Analysis of HSF1 structural and functional domains: a model for the mechanism of activation of HSF1. (A) Schematic representation of HSF1 structural motifs [DNA-binding domain, heptad repeat A/B (HR-A/B) and HR-C] and functional domains (negative regulation domain, activation domain, constitutive phosphorylation, and suggested inducible phosphorylation). The location of each structural motif or regulatory domain is indicated by the amino acid position. (B) Model for HSF1 regulation. In control cells, intramolecular interactions between HR-A/B and HR-C are important for negative regulation. Additional interactions between the negative regulatory domain and the activation domain are also indicated. During heat shock and other stresses, HSF1 activation is mediated by disruption of the intramolecular interactions between HR-A/B and HR-C and between the negative regulatory domain and the activation domain in favour of new intermolecular interactions of the HR-A/B domains, leading to HSF1 trimerization and subsequent acquisition of DNA-binding activity. Reproduced from Morimoto, R.I., et al. (1997) Essays Biochem. **32**, 17–29 (Portland Press, London), with the permission of the Biochemical Society.

the extreme C-terminal 108 amino acids, corresponding to a region rich in acidic and hydrophobic residues. The C-terminal trans-activation region, corresponding to AD2, is likely to represent a large domain, as loss of residues 395–425 or 451–503, located at either end of this activation domain, severely diminished activity. The minimal activation domain of HSF1 is also sufficient to generate an enhanced (4-fold) transcriptional response to heat shock or cadmium treatment. These results demonstrate that the transcriptional activation domain of HSF1 is maintained in a negatively regulated state, presumably through intramolecular interactions, and that the signal for stress induction is also mediated by the C-terminal activator region of HSF1.

Role of HSF phosphorylation

Phosphorylation might be yet another mechanism by which HSF1 function is modulated. The initial suggestion that phosphorylation might be related to the activation of HSF came from studies showing a stress-induced change in the electrophoretic mobility of HSF that is reversed by phosphatase treatment [27,41–43]. These results were extended by direct ^{32}P-labelling experiments, which demonstrated increased phosphorylation at serine and threonine residues for *S. cerevisiae* HSF under conditions of heat shock [26]. However, there are indications that phosphorylation might also be involved in the de-activation of yeast HSF. Mutation of a serine-rich stretch in *K. lactis* HSF leads to a failure in HSF de-activation [44]. One interpretation of these results is that HSF function is positively or negatively regulated by phosphorylation of specific residues that determine the fate of the activity of the factor.

In mammalian cells, HSF1 is phosphorylated both constitutively and inducibly at serine residues [45,46]. Recent studies determined independently that constitutive phosphorylation of HSF1 at serine residues distal from the transcriptional activation domain acts to repress trans-activation [45–47]. Mutational analysis and tryptic phosphopeptide analysis of chimaeric GAL4–HSF1 deletion and point mutants identified a region of constitutive phosphorylation encompassing serine residues 303 and 307 [46]. In these studies, a transfected wild-type GAL4–HSF1 chimaera is repressed for transcriptional activity and de-repressed by heat shock. However, mutation of serines 303 and 307 to alanine results in de-repression to give high levels of constitutive activity [46], and mutation of the same residues to glutamic acid, which mimics a phosphorylated serine, results in normal repression at control temperature and normal heat-shock inducibility [47]. Hence these studies revealed that constitutive phosphorylation plays an important role in the negative regulation of HSF1 transcriptional activity at control temperatures.

In contrast, the effects of inducible phosphorylation on the activity of HSF1 are not as well understood. Anti-inflammatory drugs have been used to study the role of inducible phosphorylation on the activity of HSF1 [45]. In human cells, treatment with either salicylate or indomethacin induces the acquisition of HSF1 DNA-binding activity to levels comparable with those observed during heat shock, yet the drug-induced form of HSF1 is transcriptionally inert [48,49]. Comparison of the heat-shock- and drug-induced forms of HSF1 revealed that the transcriptionally inert drug-induced HSF1 is constitutively but not inducibly serine-phosphorylated, whereas heat-shock-induced HSF1 is both constitutively and inducibly serine-phosphorylated. Results fom these studies demonstrated that the heat-shock transcriptional response is a multi-step process in which trimerization and acquisition of DNA-binding activity by HSF1 are necessary but insufficient for transcriptional activation. Furthermore, the data also show that acquisition of the trimeric DNA-binding state of HSF1 is independent of and precedes inducible phosphorylation, and that inducible phosphorylation correlates with transcriptional activation.

Cell biology of HSF1 activation

Studies on the subcellular localization of HSF1 have provided additional insights into the mechanism of regulation of the heat-shock response. Analyses of the localization of HSF1 in unstressed cells have detected the inert protein in the cytosol [27,29,34,41,50–53], in both nuclear and cytosolic compartments [27], or primarily in the nucleus [34,52]. These differences have been attributed primarily to the recognition properties of the various antibodies, which appear to detect distinct cytosolic and nuclear populations of HSF1 [16,52].

No discrepancies have been reported with regard to the subcellular localization of the heat-shocked form of HSF1, which is localized primarily in the nucleus of the cell. Furthermore, upon exposure of HeLa cells to heat shock, HSF1 relocalizes to form brightly staining nuclear foci or granules, which have been detected by immunofluorescence with polyclonal and monoclonal antibodies [27,54,55]. Within 30 min of heat shock, HSF1 granules ranging in size from 0.5 to 1.5 μm can be detected in 80–90% of HeLa cells (Fig. 4A); by 60 min of heat shock, 95% of the cells exhibit brightly staining HSF1 granules. Up to 2 h of heat shock, HSF1 granules are ubiquitous in all cells. Analysis of the size distribution of the granules in HeLa cells heat-shocked at 42°C for 2 h reveals that they can be described as two populations. Thus 60% of the foci correspond to smaller (0.5–1.5 μm) brightly staining granules or speckles; those remaining are larger (1.5–2.5 μm) clustered or ring-like granular structures. The majority (55%) of the cells contained an average of seven HSF1 granules, although substantial cell-to-cell variation can be detected. Comparison with the level of HSF1 DNA-binding activity (Fig. 4B) reveals that the appearance of HSF1 granules correlates closely with both the acquisition of HSF1 DNA-binding activity and the inducibly phosphorylated state of HSF1 (Figs. 4B and 4C). After 2 h of continuous heat shock, the fraction of cells which exhibit HSF1 granules declines rapidly to 5% of the population; likewise, HSF1 DNA binding attenuates and it is dephosphorylated to the control state (Figs. 4B and 4C). These results suggest that HSF1 granules represent a new component of the heat-shock response in human cells, as the temporal link in the well-established kinetics of the heat-shock response and the rapid recovery during attenuation parallel precisely the transient appearance and disappearance of HSF1 granules.

Although the presence of HSF1 granules appears to be a reliable visual indicator of the transcriptional activity of HSF1, the precise role of these foci remains uncertain. Results from experiments designed to determine whether HSF1 granules represent a novel nuclear domain or correspond to previously characterized sub-nuclear structures have shown that the granules do not co-localize with sites of DNA replication, RNA splicing or the nucleolus [54]. Furthermore, fluorescence *in situ* hybridization experiments have shown that HSF1 granules do not co-localize with the loci of the Hsp70 and Hsp90 genes [55]. One possibility is that HSF1 granules provide a means to locally increase the concentration of HSF1 to ensure rapid kinetics of HSF1 DNA binding and elevated transcription of heat-shock genes. At least one observation seems to support

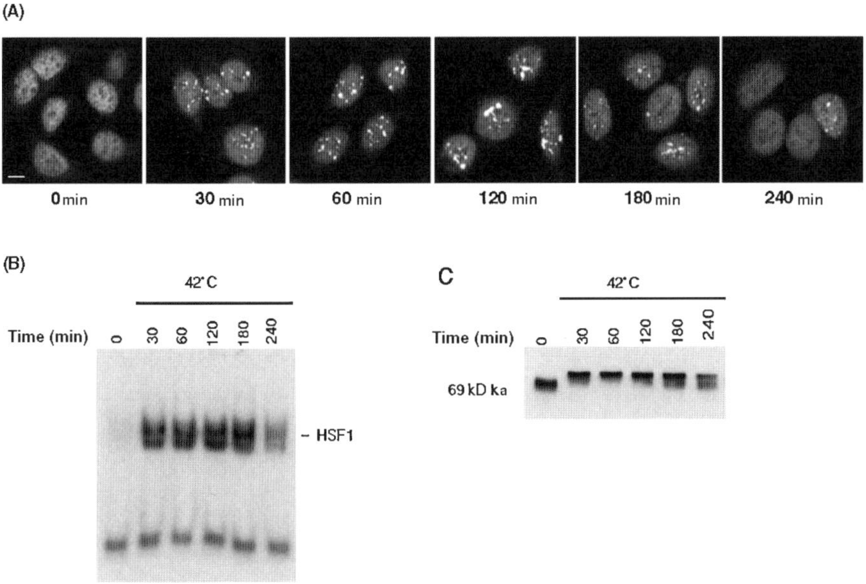

Fig. 4. Kinetics of HSF1 granule formation and comparison with DNA-binding activity and inducible phosphorylation during heat shock. (A) Immunofluorescence analysis in HeLa cells prior to heat shock (0 min) or after incubation at 42°C for 60, 120, 180 and 240 min using the HSF1-directed rat monoclonal antibody 10H8. The bar represents 5 μm. (B) Kinetics of HSF1 DNA-binding activity measured using gel mobility shift analysis. (C) Western blot analysis of whole-cell extracts from samples indicated in (B). The lower mobility of HSF1 in heat-shocked cells is due to inducible serine phosphorylation.

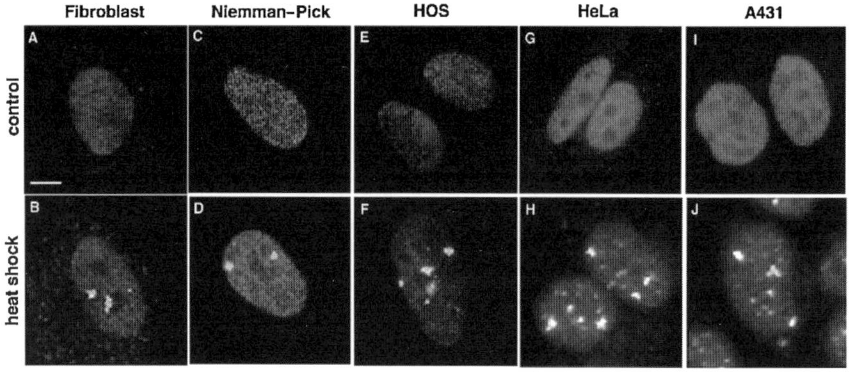

Fig. 5. Subcellular localization of HSF1 in various primary and transformed human cells. Cells either at the control temperature or at 42°C (heat-shock conditions) were analysed by immunofluorescence using rat monoclonal antibody 10H8 against HSF1. Human primary fibroblasts (A, B), epithelial cells (C, D), HOS cells (E, F), HeLa cells (G, H) and A431 cells (I, J) were examined. The bar represents 5 μm.

this hypothesis. When primary fibroblasts and epithelial cells are exposed to heat shock, brightly staining HSF1 granules can be detected, with the majority of cells containing two large foci and occasional smaller speckles (Figs. 5B and 5D). Examination of the HSF1 staining pattern in two other human transformed cell lines, HOS (hyperdiploid osteosarcoma cell line) and A431 (hypotetraploid epidermal carcinoma cell line), revealed that HOS cells have an average of four to five granules per nucleus (Fig. 5F), whereas HeLa (hypotetraploid cervical carcinoma cell line) and A431 cells contain an average of seven granules per nucleus (Figs. 5H and 5J). The number of HSF1 granules per cell is relatively constant in any particular cell line, with fewer granules being detected in primary cells than in transformed cell lines [54,55]. These results suggest a possible relationship between the number of granules and chromosomal ploidy, and the existence of specific chromosomal targets for HSF1 foci. However, further investigation will be required in order to determine the precise role that HSF1 granules have in the heat-shock response.

References

1. Ritossa, F.M. (1962) Experientia **18**, 571–573
2. Ashburner, M. (1970) Chromosoma **31**, 356–376
3. Greenleaf, A.L., Plagens, U., Jamrich, M. and Bautz, E.K.F. (1978) Chromosoma **65**, 127–136
4. Jamrich, M., Haars, R., Wulf, E. and Bautz, E.K.F. (1977) Chromosoma **64**, 319–326
5. Lindquist, S. (1986) Annu. Rev. Biochem. **55**, 1151–1191
6. Lindquist, S. and Craig, E.A. (1988) Annu. Rev. Biochem. **22**, 631–677
7. Craig, E.A., Gambill, B.D. and Nelson, R.J. (1993) Microbiol. Rev. **57**, 402–414
8. Georgopoulos, C. and Welch, W.J. (1993) Annu. Rev. Cell Biol. **9**, 601–634
9. Gething, M.J. and Sambrook, J. (1992) Nature (London) **355**, 33–45

10. Hartl, F.U. (1996) Nature (London) **381**, 571–579
11. Hendrick, J.P. and Hartl, F.U. (1993) Annu. Rev. Biochem. **62**, 349–384
12. Parsell, D.A. and Lindquist, S. (1993) Annu. Rev. Genet. **27**, 437–496
13. Edington, B.V., Whelan, S.A. and Hightower, L.E. (1989) J. Cell. Physiol. **139**, 219–228
14. Hightower, L.E. (1991) Cell **66**, 191–197
15. Pelham, H.R. (1986) Cell **46**, 959–961
16. Morimoto, R.I., Jurivich, D.A., Kroeger, P.E., Mathur, S.K., Murphy, S.P., Nakai, A., Sarge, K., Abravaya, K. and Sistonen, L.T. (1994) in The Biology of Heat Shock Proteins and Molecular Chaperones (Morimoto, R.I., Tissieres, A. and Georgopoulos, C., eds.), pp. 417–456, Cold Spring Harbor Laboratory Press, Cold Spring Harbor
17. Wu, C. (1995) Annu. Rev. Cell. Dev. Biol. **11**, 441–469
18. Nakai, A. and Morimoto, R.I. (1993) Mol. Cell. Biol. **13**, 1983–1997
19. Nakai, A., Tanabe, M., Kawazoe, Y., Inazawa, J., Morimoto, R.I. and Nagata, K. (1997) Mol. Cell. Biol. **17**, 469–481
20. Rabindran, S.K., Giorgi, G., Clos, J. and Wu, C. (1991) Proc. Natl. Acad. Sci. U.S.A. **88**, 6906–6910
21. Sarge, K.D., Zimarino, V., Holm, K., Wu, C. and Morimoto, R.I. (1991) Genes Dev. **5**, 1902–1911
22. Scharf, K.D., Rose, S., Zott, W., Schoff, F. and Nover, L. (1990) EMBO J. **9**, 4495–4501
23. Schuetz, T.J., Gallo, G.J., Sheldon, L., Tempst, P. and Kingston, R.E. (1991) Proc. Natl. Acad. Sci. U.S.A. **88**, 6911–6915
24. Chen, Y., Barlev, N.A., Westergaard, O. and Jakobsen, B.K. (1993) EMBO J. **12**, 5007–5018
25. Nieto-Sotelo, J., Wiederrecht, G., Okuda, A. and Parker, C.S. (1990) Cell **62**, 807–817
26. Sorger, P.K. (1990) Cell **62**, 793–805
27. Sarge, K.D., Murphy, S.P. and Morimoto, R.I. (1993) Mol. Cell. Biol. **13**, 1392–1407 (published errata appear in Mol. Cell. Biol. **13**, 3122–3123 and 3838–3839)
28. Sistonen, L., Sarge, K.D., Phillips, B., Abravaya, K. and Morimoto, R.I. (1992) Mol. Cell. Biol. **12**, 4104–4111
29. Sistonen, L., Sarge, K.D. and Morimoto, R.I. (1994) Mol. Cell. Biol. **14**, 2087–2099
30. Theodorakis, N.G., Zand, D.J., Kotzbauer, P.T., Williams, G.T. and Morimoto, R.I. (1989) Mol. Cell. Biol. **9**, 3166–3173
31. Fiorenza, M.T., Farkas, T., Dissing, M., Kolding, D. and Zimarino, V. (1995) Nucleic Acids Res. **23**, 467–474
32. Sarge, K.D., Park, S.O., Kirby, J.D., Mayo, K.E. and Morimoto, R.I. (1994) Biol. Rep. **50**, 1334–1343
33. Nakai, A., Kawazoe, Y., Tanabe, M., Nagata, K. and Morimoto, R.I. (1995) Mol. Cell. Biol. **15**, 5268–5278
34. Westwood, J.T., Clos, J. and Wu, C. (1991) Nature (London) **353**, 822–827
35. Westwood, J.T. and Wu, C. (1993) Mol. Cell. Biol. **13**, 3481–3486
36. Rabindran, S.K., Haroun, R.I., Clos, J., Wisniewski, J. and Wu, C. (1993) Science **259**, 230–234
37. Treuter, E., Nover, L., Ohme, K. and Scharf, K.-D. (1993) Mol. Gen. Genet. **240**, 113–125
38. Green, M., Schuetz, T.J., Sullivan, E.K. and Kingston, R.E. (1995) Mol. Cell. Biol. **15**, 3354–3362
39. Shi, Y., Kroeger, P.E. and Morimotor, R.I. (1995) Mol. Cell. Biol. **15**, 4309–4318
40. Zuo, J., Rungger, D. and Voellmy, R. (1995) Mol. Cell. Biol. **15**, 4319–4330

41. Larson, J.S., Schuetz, T.J. and Kingston, R.E. (1988) Nature (London) **335**, 372–375
42. Sorger, P.K., Lewis, M.J. and Pelham, H.R. (1987) Nature (London) **329**, 81–84
43. Sorger, P.K. and Pelham, H.R. (1988) Cell **54**, 855–864
44. Hoj, A. and Jakobsen, B.K. (1994) EMBO J. **13**, 2617–2624
45. Cotto, J.J., Kline, M. and Morimoto, R.I. (1996) J. Biol. Chem. **271**, 3355–3358
46. Kline, M.P. and Morimoto, R.I. (1997) Mol. Cell. Biol. **17**, 2107–2115
47. Knauf, U., Newton, E.M., Kyriakis, J. and Kingston, R.E. (1996) Genes Dev. **10**, 2782–2793
48. Jurivich, D.A., Sistonen, L., Kroes, R.A. and Morimoto, R.I. (1992) Science **255**, 1243–1245
49. Lee, B.S., Chen, J., Angelidis, C., Jurivich, D.A. and Morimoto, R.I. (1995) Proc. Natl. Acad. Sci. U.S.A. **92**, 7207–7211
50. Baler, R., Dahl, G. and Voellmy, R. (1993) Mol. Cell. Biol. **13**, 2486–2496
51. Mosser, D.D., Kotzbauer, P.T., Sarge, K.D. and Morimoto, R.I. (1990) Proc. Natl. Acad. Sci. U.S.A. **87**, 3748–3752
52. Wu, C., Clos, J., Giorgi, G., Haroun, R.I., Kim, S.J., Rabindran, S.K., Westwood, J.T., Wisniewski, J. and Yim, G. (1994) in The Biology of Heat Shock Proteins and Molecular Chaperones (Morimoto, R.I., Tissieres, A. and Georgopoulos, C., eds.), pp. 395–416, Cold Spring Harbor Laboratory Press, Cold Spring Harbor
53. Zimarino, V., Tsai, C. and Wu, C. (1990) Mol. Cell. Biol. **10**, 752–759
54. Cotto, J.J., Fox, S.G. and Morimoto, R.I. (1997) J. Cell Sci. **110**, 2925–2934
55. Jolly, C., Morimoto, R.I., Robert-Nicoud, M. and Vourc'h, C. (1997) J. Cell Sci. **110**, 2935–2941

Biochem. Soc. Symp. **64**, 119–128
Printed in Great Britain

9

Transcriptional regulation via redox-sensitive iron–sulphur centres in an oxidative stress response

Bruce Demple*, Elena Hidalgo and Huangen Ding

Department of Cancer Cell Biology,
Harvard School of Public Health, Boston, MA 02115, U.S.A.

Abstract

Genetic responses to oxidative stress are triggered by excessive levels of agents such as superoxide. The *soxRS* regulon of *Escherichia coli* includes at least a dozen oxidative-stress and antibiotic-resistance genes that are activated by the SoxS protein, the synthesis of which is controlled by the redox-sensing SoxR protein. SoxR is a homodimer of 17 kDa subunits, each of which contains a [2Fe–2S] cluster. Transcriptional activation by SoxR is controlled by the oxidation state of these metal centres. In the absence of oxidative stress, the [2Fe–2S] centres are in the reduced form and the protein is inactive, although it still binds the *soxS* promoter. Agents that generate superoxide in the cell (e.g. paraquat) cause rapid oxidation of the metal centres, which triggers the transcriptional activity of SoxR; removal of the oxidative stress is followed by rapid re-reduction of the [2Fe–2S] centres. This facile mechanism links oxidation state to control of protein activity and may be used widely to allow cells to respond to oxidative stress.

Reactive oxygen and cellular stress responses

Aerobic organisms must cope with the toxic side-effects of the oxygen upon which they depend. This toxicity is potentiated by various reactions, with the pivotal ones generating reactive derivatives that can damage key cellular components. These inadvertent toxicants include superoxide ($O_2^{-\bullet}$), hydrogen peroxide and hydroxyl radical ($^{\bullet}OH$). Super-

*To whom correspondence should be addressed.

oxide is formed through autoxidation reactions, such as the oxidation of reduced NADH oxidase by O_2 in mitochondria; other proteins throughout the cell are prone to similar superoxide-generating autoxidation [1,2]. The immune system produces $O_2^{-\bullet}$ deliberately by activating membrane-bound NADPH oxidase, which is part of the cytotoxic weaponry arrayed against pathogens and tumour cells [3]. Immune cells also produce large amounts of another reactive species, nitric oxide ($NO^\bullet$) [4].

The primary products $O_2^{-\bullet}$ and $NO^\bullet$ can both disrupt protein iron–sulphur centres, although the details of the underlying chemistry are different and poorly understood [5–7]. Superoxide can oxidize and disrupt protein [4Fe–4S] centres to inactivate enzymes such as aconitase and release their iron into the cell. The released iron can participate in other damaging reactions [8] by reacting with H_2O_2 to generate $^\bullet OH$, which is capable of damaging all the major macromolecules of the cell. $NO^\bullet$ and $O_2^{-\bullet}$ combine avidly to form still another compound, peroxynitrite ($ONOO^-$), which has reactivity similar to that of $^\bullet OH$ but can also destroy iron–sulphur centres [9].

The threat of reactive products of oxygen and nitrogen has led to the evolution of various antioxidant defence systems [10–12]. Superoxide dismutase (SOD) limits $O_2^{-\bullet}$ to acceptable levels (approx. 10^{-11} M), and in the process generates H_2O_2. Catalase, in turn, limits the steady-state concentration of H_2O_2 to approx. 10^{-7} M [13]. Other enzymic and non-enzymic functions prevent or correct macromolecular damage caused by oxidative agents. These defences are regulated in many organisms at the level of gene expression, and the induction of key antioxidant enzymes often fits the expectation for an adaptive response [14]. Thus SOD may be induced by agents such as paraquat, which generates $O_2^{-\bullet}$ intracellularly; increased expression of SOD then limits the accumulation of $O_2^{-\bullet}$. Catalase is frequently found to be inducible by H_2O_2. In each case, the 'stress' of excess oxidant is countered by an enzyme that rids the cell of that agent.

During the past decade and more, the induction of antioxidant enzymes has been revealed as a facet of more complex genetic responses. While such co-ordinate control is beginning to emerge as a theme in eukaryotes, studies in bacteria have provided the most well-defined examples of multi-gene responses to oxidative stress [14]. Two systems are known for *Escherichia coli*: the *oxyR* regulon consists of eight or nine genes that are induced by the H_2O_2-responsive OxyR protein, and the *soxRS* regulon is under the two-stage control of SoxS and the superoxide-responsive SoxR protein.

The *soxRS* regulon: a two-stage system of gene activation

The *soxRS* regulon co-ordinates the expression of at least 12 genes in response to $O_2^{-\bullet}$-generating agents such as paraquat, or to nitric oxide [14]. Many of these genes have evident roles in combating oxidative stress:

sodA encodes Mn-containing SOD; *zwf* encodes glucose-6-phosphate dehydrogenase (for resupplying NADPH used up in antioxidant reactions); *fpr* encodes ferredoxin:NADPH oxidoreductase to restore reduced iron–sulphur centres; *fumC* encodes fumarase and *acnA* encodes aconitase to replace enzymes inactivated by oxidation; *nfo* encodes a DNA repair enzyme that acts specifically on oxidative damage. However, activation of the *soxRS* system also mobilizes mechanisms that mediate resistance to antibiotics (such as decreasing drug uptake [15] or increasing drug efflux [16]). The phenotypic resistances mediated by *soxRS* have recently been expanded to include organic solvents and toxic metals [17]. The evolutionary basis of the connection between oxidative stress and the various phenotypes is not clear, but it could reflect the existence of environmental redox agents against which the antibiotic- and solvent-resistance mechanisms are effective. Whatever the case, this oxidative-stress-inducible antibiotic resistance may have clinical significance, because phagocytosis of *E. coli* by NO-generating macrophages activates the *soxRS* system [18,19], which could increase the antibiotic resistance of the phagocytosed bacteria.

The *soxRS* regulon genes listed above are activated directly by the SoxS protein [20,21], a member of the XylS/AraC family of transcription activators. The SoxS polypeptide, just 106 residues in length, is homologous to the extreme C-terminus of AraC-type proteins [22,23]. A close homologue of SoxS is the *E. coli* MarA protein (42% identity [24]), and this similarity reflects a mirror-image relationship between responses to oxidative stress and antibiotics: MarA is the gene activator of the *marRAB* regulon, which is triggered by antibiotics and weak acids such as salicylate [25,26]. SoxS and MarA induce overlapping (but non-identical) sets of genes; e.g. both proteins activate *sodA* and *zwf*, but only SoxS acts on *nfo* [14]. SoxS and MarA activate transcription in a fashion typical of AraC-type proteins: they bind sites abutting or just 5′ of the −35 promoter element and enhance the subsequent binding of σ^{70}-containing RNA polymerase [21,27].

The activities of SoxS and MarA in cells are controlled by regulating the transcription of their structural genes. For SoxS, this regulation arises from a strict dependence on the activation of SoxR protein. Thus SoxR is the pivotal redox-sensing component of the *soxRS* regulatory pathway. As the cellular concentration of SoxS rises, target promoters of the regulon become activated, which finally leads to induction of the various proteins that mediate resistance to oxidative stress and antibiotics (Fig. 1).

SoxR: a transcription activator with iron–sulphur centres

Genetic demonstration that SoxR is the fundamental redox-sensing component of the *soxRS* system led to a focus on the properties of this protein. The finding that SoxR in crude extracts binds the *soxS* promoter [28] indicated a direct transcriptional role and provided a functional assay

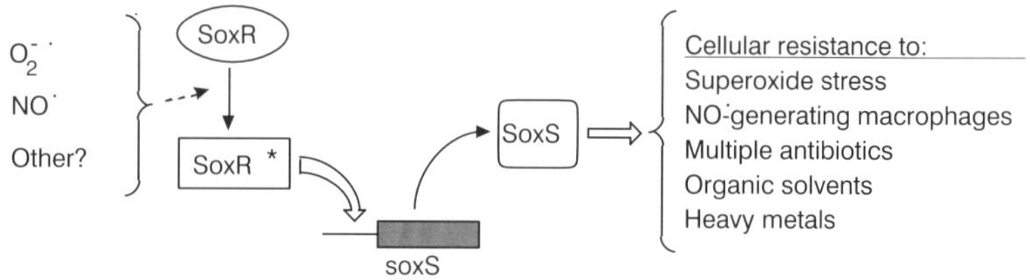

Fig. 1. Regulatory organization and phenotypes of the *soxRS* regulon. An intracellular redox signal converts existing SoxR protein into an active form that stimulates expression of the *soxS* gene. Synthesis of SoxS protein then leads to activation of genes that confer resistance to oxidants, antibiotics, solvents and some metals. Reproduced with permission from Hidalgo, E., Ding, H. and Demple, B. (1997) Redox signal transduction via iron–sulfur clusters in the SoxR transcription activator. *Trends Biochem. Sci.* **22(6)**, 207–210.

for monitoring the protein during its isolation. The initial purification [29] quickly revealed an important physical property of SoxR: the purified protein has a distinct reddish-brown colour suggestive of the presence of metals or other non-protein functional groups. Analysis for various metals revealed the only significant element to be iron [29], and indeed the visible absorption spectrum of SoxR is similar to that of iron–sulphur proteins, particularly those with binuclear metal centres ([2Fe–2S]). Consistent with this interpretation, inorganic sulphide was also present in purified SoxR at a stoichiometry close to $2S^{2-}$ per protein monomer. EPR spectroscopy, which detects the unpaired electrons of paramagnetic species such as Fe^{2+}, ultimately confirmed the presence of a pair of [2Fe–2S] centres in a dimeric SoxR protein [30,31].

A key question was apparent early: what is the role of the [2Fe–2S] centres in SoxR? As it happens, purification of SoxR in the presence of monomeric thiols such as β-mercaptoethanol strips off the Fe from the protein [29]; we now know that this loss is due to a complex series of oxygen-dependent reactions in which free radicals, perhaps thiol-based, destroy the [2Fe–2S] centres [32]. To our surprise, the iron-free apo-SoxR protein bound the *soxS* promoter with an affinity indistinguishable from that of Fe-SoxR [29]. However, only Fe-SoxR was able to activate *in vitro* transcription from a *soxS* promoter fragment [29,32]. The metal centres of SoxR protein are therefore essential for its transcriptional activity, independent of DNA binding, and they seemed a promising focus for studies of the redox chemistry that activates the *soxRS* system.

Allosteric transcriptional activation by SoxR

Before continuing a discussion of the regulation of the activity of SoxR, it is worth considering briefly the current picture of the transcription-activating mechanism of this protein. As mentioned above, SoxR has the unusual property that both active and inactive forms bind the DNA target site equally well. An important clue to the mechanism was the identification of the SoxR binding site by footprinting analysis. These studies revealed that SoxR binds a perfectly symmetrical sequence that lies exactly between the −35 and −10 promoter elements [29,30]. Such a site would ordinarily be associated with transcriptional repression, but there is no indication of a repressor action by SoxR at the wild-type *soxS* promoter [28,33,34].

Details of another protein with similar properties were emerging at that time: MerR, which governs mercury-resistance operons of some plasmids and transposons [35]. MerR is thought to compensate for suboptimal spacing in the *mer* promoter, in which the −35 and −10 elements are separated by 19 bp rather the optimal 17 ± 1 bp. MerR also binds its target promoter in both the activated and non-activated states [35]. MerR protein is in fact homologous to SoxR: these ~17 kDa polypeptides have 28% identity [20]. MerR has been proposed to twist and unbend the *mer* promoter to mediate transcriptional activation [36].

Such a 'promoter remodelling' mechanism may apply to SoxR as well. The position of the *soxS* transcription start site and the positions of probable −10 and −35 hexamers in the upstream DNA sequence are consistent with a 19 bp spacer in this promoter [29]. Recent data [34] support this conclusion dramatically: mutant *soxS* promoters engineered to have 16–18 bp spacing drive strong, SoxR-independent transcription *in vivo*, and derivatives with 16 or 17 bp spacers are strongly repressed by SoxR binding. In the repressible *soxS* promoter mutants, SoxR binding *in vitro* interferes with binding by RNA polymerase [34].

Redox-regulated transcription by SoxR *in vitro* and *in vivo*

Our initial hypothesis was that the transcriptional activity of SoxR is controlled through oxidation and reduction of the [2Fe–2S] centres [29]. This hypothesis was attractive in suggesting a facile way to link redox reactions with transcriptional control. Testing this idea proved technically difficult, however, and the critical experiments were performed only recently (see below). In the meanwhile, alternative and more experimentally tractable hypotheses were tested. One possibility was that SoxR activity is controlled through assembly and disassembly of the [2Fe–2S] centres. The existence of stable apo-SoxR [29], its generation by the biological thiol glutathione [32] and its ability to bind the *soxS* promoter were consistent with such a possibility. Moreover, transcriptionally active SoxR could be rapidly generated from the apoprotein *in vitro* by reconstitution of the [2Fe–2S] centres in a reaction facilitated by *Azotobacter vinelandii* NifS protein [37], which generates S^{2-} from L-cysteine [38]. A serious problem with this idea, however, was that even cells not treated to activate SoxR yielded the iron-containing protein with full transcriptional activity [29,39]. It is much easier to imagine that protein isolation leads to fortuitous oxidation, rather than the accidental re-assembly of the centres into apoprotein. Nonetheless, under some circumstances yet to be discovered there may be a role for regulation of SoxR activity through its iron content.

A turning point came with the development of *in vitro* transcription systems in which the redox potential could be carefully controlled. Titration with the chemical reductant dithionite showed that the SoxR [2Fe–2S] centres have a midpoint redox potential of about −285 mV [39,40], quite close to values reported for the redox potential in living cells. The standard (aerobic) transcription assays for SoxR had been carried out at a redox potential of about +200 mV; when this was reduced anaerobically to −350 mV, transcriptional activity was lost, whereas re-oxidation to +100 mV restored transcription. This oxidation–reduction process could be repeated many times [39,40], which demonstrated that the SoxR [2Fe–2S] centres act as redox-regulated transcriptional switches *in vitro* (Fig. 2). Weiss and colleagues showed that the interconversion of oxidized and reduced SoxR did not significantly affect the DNA-binding affinity of the protein [39], consistent with the independence of binding from the presence of the [2Fe–2S] centres [29].

In vivo evidence strongly supports the *in vitro* experiments. In cells overexpressing wild-type SoxR protein, the EPR signal characteristic of its reduced [2Fe–2S] centres is observed with an intensity corresponding to

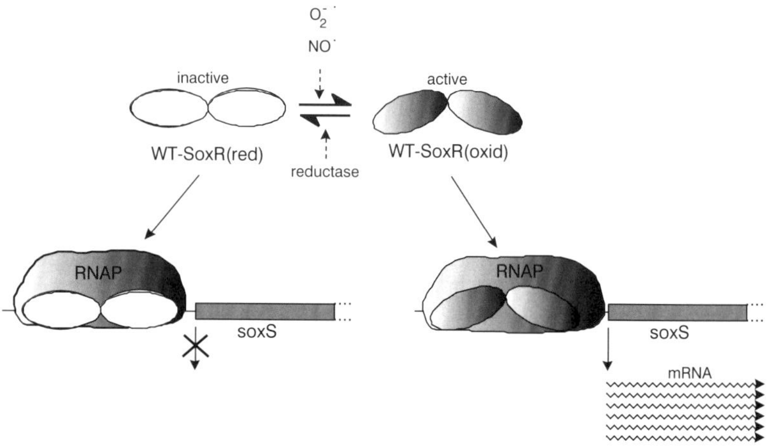

Fig. 2. Allosteric activation of SoxR. SoxR with reduced [2Fe–2S] centres [WT-SoxR(red)], probably maintained by reductase activity, is shifted to the oxidized form [WT-SoxR(oxid)] in cells exposed to nitric oxide or superoxide-generating agents. Both forms of SoxR bind the *soxS* promoter and do not interfere with binding by σ^{70}-containing RNA polymerase (RNAP), but only oxidized SoxR activates transcription. Reproduced with permission from Hidalgo, E., Ding, H. and Demple, B. (1997) Redox signal transduction via iron–sulfur clusters in the SoxR transcription activator. *Trends Biochem. Sci.* **22(6)**, 207–210.

≥95% of the protein existing as reduced Fe-SoxR [41,42]. The EPR signal is eliminated in cells treated with superoxide-generating agents and restored after the agents are removed or oxygenation is stopped [41,42]. In contrast, several constitutively active mutant forms of SoxR fail to accumulate with reduced [2Fe–2S] centres even during normal growth [42,43]. These proteins still contain the iron–sulphur centres, and are dependent on them in the oxidized state for transcriptional activity [43]. At least one mutant protein, with a single amino acid substitution in the centre of the polypeptide, has a redox midpoint potential shifted by −65 mV [43]; such a change could be sufficient to shift the *in vivo* redox equilibrium of this protein to the oxidized form.

Two other constitutively active SoxR proteins have apparently normal midpoint potentials [43]. The alteration in one of these proteins is near the C-terminus, just beyond a cluster of four cysteine residues that anchors the [2Fe–2S] centres. Deletions into this region that fall short of the cysteines render SoxR constitutively active [44], and several such engineered mutant proteins also failed to remain reduced *in vivo* [42]. One speculation is that the SoxR C-terminus constitutes a site for interaction with a reductase that maintains the protein in the reduced form [42–44]. An attractive candidate for this enzyme might be the *soxRS*-regulated ferredoxin:NADPH oxidoreductase [14], but this enzyme was reported not to act on SoxR *in vitro* [39].

Wild-type SoxR *in vivo* is subject to rapid oxidation and reduction. EPR measurements on direct culture samples showed that the [2Fe–2S] centres are converted into the oxidized form (≥95%) within 2 min after exposure to paraquat or other superoxide-generating agents [41]. When active aeration was stopped, the reduced centres were restored in just a few minutes. This rapid switching of the SoxR oxidation state is paralleled exactly by its transcriptional activity *in vivo*: the level of *soxS* mRNA increases approx. 20-fold within 2 min after paraquat exposure, and within 10 min it rises to a new steady state ~100-fold higher than the basal level [41].

The attainment of a new steady state so soon after induction is initiated, in the continuing presence of activated SoxR protein, suggested that there might be other limits to the accumulation of *soxS* mRNA. Indeed this proved to be the case: the chemical half-life of the *soxS* message decreases significantly during oxidative stress, from 3.6 min to 1.2 min [41]. This observation indicates a new level of control in response to oxidative stress, but the mechanism of this change remains unknown.

Iron–sulphur centres and redox signalling

The activation of *soxS* transcription by SoxR seems to result from a structural alteration to the promoter DNA. This change must, in turn, depend on changes in SoxR structure that are transmitted in the protein–DNA complex, i.e. an allosteric transition. The mechanism by which changes in the oxidation state of [2Fe–2S] centres might drive this transition is unknown, and large structural changes in iron–sulphur geometry are not commonly associated with changes in the oxidation state. One exception is the [4Fe–4S] centre of nitrogenase that is allosterically linked to a distant ATPase site [45,46]. In that case, the metallocentre is positioned between the subunits of a heterodimer, which might facilitate the structural change. In the *Pseudomonas putida* putidaredoxin, the oxidation state of a [2Fe–2S] centre dramatically influences binding to a cytochrome *P*-450 enzyme [47], but the underlying structural changes are unknown. While SoxR might undergo a change in the geometry of its metal centre, as in nitrogenase, an alternative simple idea is that the change in net charge upon oxidation/reduction drives a SoxR conformational transition. The redox titration of the SoxR [2Fe–2S] centres is not accurately modelled by the Nernst equation with $n=1$ [40], as though reduction of one subunit slightly influences the midpoint potential of the other. Such co-operativity or anti-co-operativity would be consistent with either a charge-driven or a co-ordination-driven mechanism of allosteric change in SoxR.

Another transcription activator is emerging as a model for redox-responsive regulation by protein iron–sulphur centres. The *E. coli* Fnr protein operates during anaerobic growth to stimulate expression of appropriate metabolic genes, and is rapidly inactivated upon the transition to aerobic growth [48,49]. The wild-type protein proved difficult to isolate in an active form, but Kiley and co-workers took advantage of mutant forms that remain active during aerobic growth [50]. This approach led to the conclusion that active Fnr contains an iron–sulphur centre of the [4Fe–4S] type, and that the metal centre is destroyed during exposure to oxygen [51]. A recent report [52] refined this model by showing that, upon exposure to O_2, the [4Fe–4S] centres of active Fnr are first converted into [2Fe–2S] centres, with the loss of transcriptional activity. Continued oxygen exposure leads to elimination of Fe from the protein, which can be restored in a reaction with the NifS protein [52], as found for the SoxR [2Fe–2S] centres.

The examples of SoxR and Fnr yield some general points. First, the appearance of iron–sulphur-dependent transcription regulatory mechanisms at least twice in one organism suggests that such metal centres may be used widely to sense changes in oxidative conditions. Such sensors need not be restricted to transcription proteins, but could in principle be harnessed to control enzymes, transport proteins, etc. Secondly, the binuclear ([2Fe–2S]) and tetranuclear ([4Fe–4S]) centres seem to have divergent properties as redox sensors. The [2Fe–2S] centres seem to be retained under oxidative conditions: the binuclear centres of SoxR are functional during oxidative stress, and Fnr converts into binuclear centres during oxygen exposure. In contrast, the tetranuclear centres of Fnr are unstable to oxygen. Hydrolyases with [4Fe–4S] centres, such as aconitase [9], and ferredoxins with tetranuclear iron–sulphur centres are also inactivated by exposure to oxygen. Spinach dihydroxy acid dehydratase, however, has [2Fe–2S] centres and is quite stable to oxygen [53]. One could therefore speculate that binuclear centres function primarily under oxidative conditions and tetranuclear centres under reducing conditions. The identification of additional iron–sulphur proteins with redox-regulated activity is needed in order to address this possibility.

We are indebted to our colleagues here and elsewhere for many stimulating discussions on the topic of redox-regulated gene expression. Work in the authors' laboratory was supported by research grants to B.D. from the U.S. National Institutes of Health (CA37831) and the Amyotrophic Lateral Sclerosis Association, and by fellowships to E.H. from the Catalan Government (CIRIT) and to H.D. from the U.S. Public Health Service (NRSA F32-ES05726).

References

1. Chance, B., Sies, H. and Boveris, A. (1979) Physiol. Rev. **59**, 527–605
2. Gonzalez-Flecha, B. and Demple, B. (1995) J. Biol. Chem. **270**, 13681–13687
3. Babior, B.M. (1992) Enzymol. Relat. Areas Mol. Biol. **65**, 49–65
4. Marletta, M.A. (1993) Adv. Exp. Med. Biol. **338**, 281–284
5. Drapier, J.-C. (1997) Methods **11**, 319–329
6. Gardner, P.R. and Fridovich, I. (1992) J. Biol. Chem. **267**, 8757–8763
7. Flint, D.H., Smyk-Randall, E., Tuminello, J.F., Draczynska-Lusiak, B. and Brown, O.R. (1993) J. Biol. Chem. **268**, 25547–25552
8. Keyer, K. and Imlay, J.A. (1996) Proc. Natl. Acad. Sci. U.S.A. **93**, 13635–13640
9. Hausladen, A. and Fridovich, I. (1994) J. Biol. Chem. **269**, 29405–29408
10. Fridovich, I. (1995) Annu. Rev. Biochem. **64**, 97–112
11. Fridovich, I. and Freeman, B. (1986) Annu. Rev. Physiol. **48**, 693–702
12. Demple, B. and Harrison, L. (1994) Annu. Rev. Biochem. **63**, 915–948
13. Gonzalez-Flecha, B. and Demple, B. (1997) J. Bacteriol. **179**, 382–388
14. Hidalgo, E. and Demple, B. (1996) in Regulation of Gene Expression in *Escherichia coli* (Lin, E.C.C. and Lynch, A.S., eds.), pp. 435–452, R.G. Landes, Austin, TX
15. Chou, J.H., Greenberg, J.T. and Demple, B. (1993) J. Bacteriol. **175**, 1026–1031
16. Ma, D., Alberti, M., Lynch, C., Nikaido, H. and Hearst, J.E. (1996) Mol. Microbiol. **19**, 101–112
17. Nakajima, H.K., Kobayashi, H., Aono, R. and Horikoshi, K. (1995) Biosci. Biotechnol. Biochem. **59**, 1323–1325
18. Nunoshiba, T., deRojas-Walker, T., Wishnok, J.S., Tannenbaum, S.R. and Demple, B. (1993) Proc. Natl. Acad. Sci. U.S.A. **90**, 9993–9997
19. Nunoshiba, T., deRojas-Walker, T., Tannenbaum, S.R. and Demple, B. (1995) Infect. Immun. **63**, 794–798
20. Amabile-Cuevas, C.F. and Demple, B. (1991) Nucleic Acids Res. **19**, 4479–4484
21. Li, Z. and Demple, B. (1994) J. Biol. Chem. **269**, 18371–18377
22. Wu, J. and Weiss, B. (1991) J. Bacteriol. **173**, 2864–2871
23. Demple, B. and Amabile-Cuevas, C.F. (1991) Cell **67**, 837–839
24. Cohen, S.P., Hächler, H. and Levy, S.B. (1993) J. Bacteriol. **175**, 1484–1492
25. Ariza, R.R., Cohen, S.P., Bachhawat, N., Levy, S.B. and Demple, B. (1994) J. Bacteriol. **176**, 143–148
26. Miller, P.F. and Sulavik, M.C. (1996) Mol. Microbiol. **21**, 441–448
27. Jair, K.W., Martin, R.G., Rosner, J.L., Fujita, N., Ishihama, A. and Wolf, R.E. (1995) J. Bacteriol. **177**, 7100–7104

28. Nunoshiba, T., Hidalgo, E., Amabile Cuevas, C.F. and Demple, B. (1992) J. Bacteriol. **174**, 6054–6060
29. Hidalgo, E. and Demple, B. (1994) EMBO J. **13**, 138–146
30. Hidalgo, E., Bollinger, Jr., J.M., Bradley, T.M., Walsh, C.T. and Demple, B. (1995) J. Biol. Chem. **270**, 20908–20914
31. Wu, J., Dunham, W.R. and Weiss, B. (1995) J. Biol. Chem. **270**, 10323–10327
32. Ding, H. and Demple, B. (1996) Proc. Natl. Acad. Sci. U.S.A. **93**, 9449–9453
33. Wu, J. and Weiss, B. (1992) J. Bacteriol. **174**, 3915–3920
34. Hidalgo, E. and Demple, B. (1997) EMBO J. **16**, 1056–1065
35. Summers, A.O. (1992) J. Bacteriol. **174**, 3097–3101
36. Ansari, A.Z., Bradner, J.E. and O'Halloran, T.V. (1995) Nature (London) **374**, 371–375
37. Hidalgo, E. and Demple, B. (1996) J. Biol. Chem. **271**, 7269–7272
38. Zheng, L. and Dean, D.R. (1994) J. Biol. Chem. **269**, 18723–18726
39. Gaudu, P. and Weiss, B. (1996) Proc. Natl. Acad. Sci. U.S.A. **93**, 10094–10098
40. Ding, H., Hidalgo, E. and Demple, B. (1996) J. Biol. Chem. **271**, 33173–33175
41. Ding, H. and Demple, B. (1997) Proc. Natl. Acad. Sci. U.S.A. **94**, 8445–8449
42. Gaudu, P., Moon, N. and Weiss, B. (1997) J. Biol. Chem. **272**, 5082–5086
43. Hidalgo, E., Ding, H. and Demple, B. (1997) Cell **88**, 121–129
44. Nunoshiba, T. and Demple, B. (1994) Nucleic Acids Res. **22**, 2958–2962
45. Wolle, D., Dean, D.R. and Howard, J.B. (1992) Nature (London) **258**, 992–995
46. Georgiadis, M.M., Komiya, H., Chakrabarti, P., Woo, D., Kornuc, J.J. and Rees, D.C. (1992) Science **257**, 1653–1659
47. Lyons, T.A., Ratnaswamy, G. and Pochapsky, T.C. (1996) Protein Sci. **5**, 627–639
48. Beinert, H. and Kiley, P. (1996) FEBS Lett. **382**, 218–221
49. Demple, B. (1997) Methods Companion Methods Enzymol. **11**, 267–278
50. Khoroshilova, N., Beinert, H. and Kiley, P.J. (1995) Proc. Natl. Acad. Sci. U.S.A. **92**, 2499–2503
51. Lazazzera, B.A., Beinert, H., Khoroshilova, N., Kennedy, M.C. and Kiley, P.J. (1996) J. Biol. Chem. **271**, 2762–2768
52. Khoroshilova, N., Popescu, C., Münck, E., Beinert, H. and Kiley, P.J. (1997) Proc. Natl. Acad. Sci. U.S.A. **94**, 6087–6092
53. Flint, D.H., Tuminello, J.F. and Emptage, M.H. (1993) J. Biol. Chem. **268**, 22369–22376

Biochem. Soc. Symp. **64**, 129–139
Printed in Great Britain

10

Adaptive responses to environmental chemicals

C.R. Wolf*||, G. Smith†, A.G. Smith‡, K. Brown§ and C.J. Henderson*

*Imperial Cancer Research Fund Molecular Pharmacology Unit, Biomedical Research Centre, University of Dundee, Ninewells Hospital and Medical School, Dundee DD1 9SY, U.K., †Biomedical Research Centre, University of Dundee, Ninewells Hospital and Medical School, Dundee DD1 9SY, U.K., ‡Centre for Genome Research, University of Edinburgh, King's Buildings, West Mains Road, Edinburgh EH9 3JQ, U.K., and §Molecular Carcinogenesis Group, CRC Beatson Laboratories, Department of Medical Oncology, Alexander Stone Building, Switchback Road, Garscube Estate, Glasgow G61 1BD, U.K.

Abstract

Adaption to chemical agents in the environment is a fundamental part of the evolutionary process and a large number of genes have evolved to specifically detoxify potentially harmful chemical agents. These genes can act at various levels within cells and determine circulating chemical or toxin concentrations, and uptake and efflux rates as well as intracellular detoxification enzymes, such as the cytochrome *P*-450-dependent monooxygenases and glutathione S-transferases. These multigene families of proteins play a central role in chemical and drug detoxification and their polymorphic expression may well be a factor in disease susceptibility.

Introduction

Numerous organisms, particularly plants and micro-organisms, produce toxic chemicals as a defence against predators or in order to compete for nutrient sources. Therefore, as part of the evolutionary process, a complex spectrum of metabolic processes has arisen to combat their toxic effects [1]. Cytoprotection can be afforded by a range of different mechanisms, including those which affect drug accumulation by

||To whom correspondence should be addressed.

preventing uptake or by accelerating efflux [2]. This is epitomized by the ATP-dependent transporters such as the multidrug-resistance transporters and the multidrug-resistance-related proteins. The next level of cellular defence is intracellular detoxification. A large number of genes have evolved with the specific purpose of detoxifying toxic chemical agents [3]. The third level of defence lies in the detection and repair of chemical-induced damage. This includes proteins involved in DNA repair or DNA-damage responses, such as p53 (see other chapters in this volume). In multicellular organisms, protection against chemical agents is more complex because toxicity is also affected by the circulating level of the toxin. This is determined by, for example, uptake of the chemical through the gastrointestinal tract and hepatic metabolism by proteins such as the cytochrome *P*-450-dependent mono-oxygenases.

Many cellular responses to chemicals are an adaptive response to toxic challenge, where exposure to specific agents results in the activation of genes that generate resistance. Some of these responses are similar to those associated with exposure to other types of stress, such as DNA damage and oxidative stress [4], whereas others are quite distinct. When normal cellular defence systems are overwhelmed, chemical toxicity and mutagenicity occur. These processes are fundamental to the pathogenesis of many human diseases, such as cancer, neurological disease and heart disease. Indeed, there are very few human diseases for which exposure to environmental agents is not thought to be an important part of the disease process. There are, therefore, many genes involved in cellular protection against toxic agents. The focus of the present article will be on those that are specifically involved in the metabolism and disposition of foreign chemicals.

Over the last thousand million years a large number of genes have evolved whose specific role is the metabolism, disposition and detoxification of environmental chemicals. In addition to their central role in the metabolism of such agents, the products of these genes also play a pivotal role in the metabolism, disposition and mechanism of action of therapeutic drugs. As such, they have been trivially termed 'drug-metabolizing enzymes'. The intracellular levels of certain of these enzymes in tumour cells have also been associated with drug resistance [5,6].

The primary role of these enzyme systems is the metabolism of absorbed lipophilic chemicals into products that are more water-soluble and can be excreted. An inability to metabolize such compounds results in their accumulation. It is well documented that environmental chemicals inert to metabolism can be stored for many years in fat tissue. The enzymes involved in foreign compound metabolism have been divided trivially into two groups, termed phase I and phase II metabolizing enzymes. Phase I enzymes such as the cytochrome *P*-450-dependent mono-oxygenases are involved in oxidation reactions at C–H bonds or at hetero atoms such as nitrogen and sulphur [7]. Phase II enzymes are involved in the conjugation of the primary oxidation products to cofactors such as glutathione, sulphate or glucuronic acid. This further increases the hydrophilicity of the chemical and, as a consequence, facilitates its excretion.

Cytochrome *P*-450 mono-oxygenases

The cytochrome *P*-450s are haemoproteins with molecular masses of approx. 50–60 kDa. Those involved in foreign compound metabolism are membrane-bound in the endoplasmic reticulum. They are termed cytochrome *P*-450s because, when reduced, they can bind carbon monoxide to produce a Soret maximum in the visible absorption spectrum at 450 nm. Cytochrome *P*-450s are terminal electron acceptors, receiving electrons from NADPH via the flavoprotein cytochrome *P*-450 reductase. Cytochrome *P*-450s, when reduced, bind molecular oxygen. In a two-electron reduction process, one atom of molecular oxygen is incorporated into the substrate and the other into a molecule of water.

The specificity of the cytochrome *P*-450 mono-oxygenases has been exploited in a wide range of important biological processes. In mammals these enzymes, in addition to their role in foreign compound metabolism, are responsible for the biosynthesis of steroid hormones, for the breakdown of cholesterol to bile acids and for the metabolism of hormones such as retinoic acid and vitamin D. In humans there may be as many as 100 cytochrome *P*-450 genes [8]. The cytochrome *P*-450s involved in foreign compound metabolism have diverged into multigene families, with each individual enzyme exhibiting a unique, yet overlapping, substrate specificity. This is probably the result of the vast number of chemical agents that are substrates for these enzymes. Cytochrome *P*-450 gene families and subfamilies are scattered throughout the human genome.

The role of the cytochrome *P*-450 mono-oxygenases in the metabolism and disposition of therapeutic drugs demonstrates unequivocally their importance in determining our responses to environmental chemical agents [9]. Intriguingly, although approximately 20 cytochrome *P*-450s involved in foreign compound metabolism have been identified, certain of these play the predominant role in the metabolism and disposition of therapeutic agents. Of particular interest and importance are cytochrome *P*-450s in the *CYP2C*, *CYP2D* and *CYP3A* gene families. Cytochrome *P*-450 CYP3A4 probably represents approx. 30% of the total hepatic *P*-450 content, yet is responsible for the metabolism of approx. 50% of therapeutic drugs (Fig. 1a). More interestingly, cytochrome *P*-450 CYP2D6 probably only represents 1–2% of hepatic cytochrome *P*-450 content [10], yet appears to be responsible for the metabolism of up to 25% of therapeutic drugs (Fig. 1b; Table 1). Intriguingly, many of the drugs metabolized by this enzyme are active in the central nervous system, indicating that this gene may have evolved specifically to protect mammals from environmental neurotoxins. Some evidence in support of this possibility is discussed below.

In most cytochrome *P*-450 reactions, substrate oxidation results in increased hydrophilicity, detoxification and elimination. However, in certain cases, metabolites are formed that are much more toxic and mutagenic than the parent compound. The conversion of exogenous chemicals into such mutagenic products is the initiating reaction which determines the ultimate carcinogenic effects of almost all environmental carcinogens and many chemical toxins. Therefore the relative levels of cytochrome *P*-450s in a particular cell type can be an important factor in determining individual sensitivity to environmental agents.

Factors influencing variability in cytochrome *P*-450 gene expression

Similarly to many other enzymes that metabolize foreign compounds, the cytochrome *P*-450 system provides an adaptive response to chemical challenge [1]. Exposure to specific chemical agents can result in profound changes in patterns of cytochrome *P*-450 gene expression. In many cases the mechanism of regulation of these genes by chemical agents remains unclear. The best understood, however, is the regulation of cytochrome *P*-450s in the CYP1A family. In this case the foreign compound binds to a cytosolic receptor, known as the Ah receptor. This releases the receptor from Hsp90 (heat-shock protein of 90 kDa), allowing

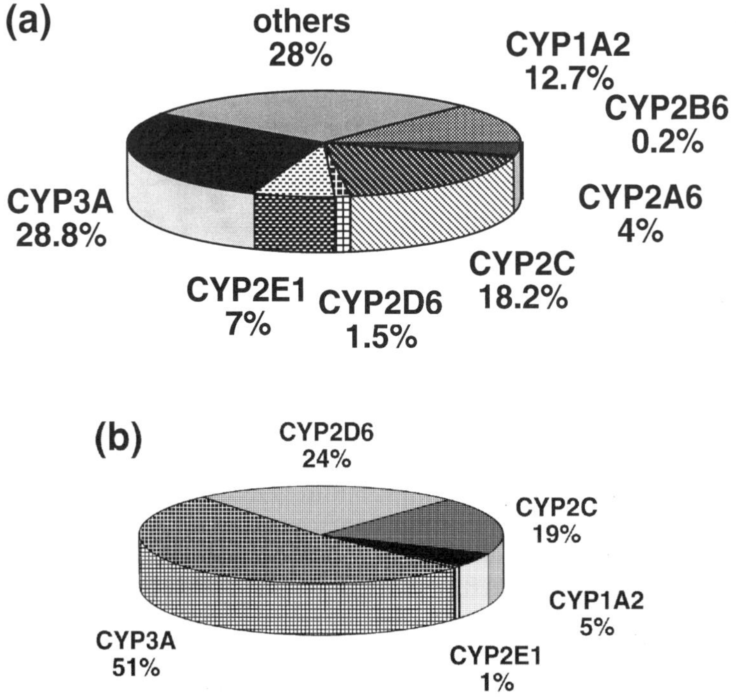

Fig. 1. Relative abundance of *P*-450 isoenzymes in human liver, and (b) relative contribution of *P*-450 isoenzymes to drug metabolism.

it to translocate to the nucleus. The Ah receptor then heterodimerizes with a second transcription factor, aryl-hydrocarbon-receptor nuclear translocator, to activate the transcription of specific cytochrome *P*-450 genes. Certain other drug-metabolizing enzymes are also regulated by this pathway. In order to fully understand how this system is regulated *in vivo*, we have generated transgenic mice in which the *CYP1A1* promoter is used to drive the expression of a reporter gene, *lacZ*. In animals not exposed to *P*-450-inducing agents, no activity from this reporter gene can be detected. However, profound changes in gene expression are observed when these animals are challenged with specific *P*-450 CYP1A1-inducing agents. Particularly marked induction is seen in tissues such as the liver, the gastrointestinal tract and the adrenal gland [11]. These studies exemplify the marked changes in gene expression that can be induced by exposure to foreign compounds. This reporter system has an interesting further application as a regulatable promoter system in transgenic animals and possibly in gene therapy [11].

Environmental and hormonal factors are obviously important in determining cellular levels of cytochrome *P*-450s. However, another important basis for individuality in *P*-450 expression is genetic polymorphism [3]. There is increasing evidence indicating that genetic

Table 1. CYP2D6 substrates.

Drug type	Examples
β-blockers	Alprenolol, metoprolol, timolol, bufuralol, propranolol, guanoxan, indoramine, bupranolol
Anti-arrhythmics	Sparteine, *N*-propylajmaline, propafenone, mexiletine, flecainide, encainide, procainamide
Tricyclic anti-depressants	Nortriptyline, desipramine, clomipramine, imipramine, amitryptiline, minaprine, fluvoxamine, luvoxamine
Anti-psychotics	Perphenazine, thioridazine, zuclopenthixol, haloperidol, tomoxetine, paroxetine, amiflavine, methoxyphenamine, fluoxetine, levomepromazine, olanzapine, perphenazine
Analgesics	Codeine, ethylmorphine
Anti-histamines	Loratadine, promethazine
Others	Debrisoquine (anti-hypertensive), 4-hydroxyamphetamine (central nervous system stimulant), phenformin, pehexiline, MDMA (ecstasy), dextromethorphan (anti-tussive), ritonavir (HIV1 protease inhibitor), dolasetron (anti-emetic), ondansetron, tropisetron (5-HT_3 receptor antagonists), nicergoline (vasodilator), mexiletine (diabetes), dexfenfluramine (appetite suppressant), 1-methyl-4-phenyl-1,2,3,6-tetrahydropyridine

polymorphism within these genes can have profound effects on drug and foreign compound disposition. Such studies are epitomized by the work on cytochrome *P*-450 CYP2D6 [9]. In the 1970s this enzyme was found to be genetically polymorphic when certain individuals involved in drug trials suffered adverse reactions due to high circulating drug concentrations, i.e. drug overdose. This overdose was ascribed to a reduced rate of drug metabolism as a consequence of the complete lack of expression of a specific cytochrome *P*-450 enzyme, CYP2D6, in affected individuals. Subsequent studies have demonstrated the presence of a variety of gene-inactivating mutations within the *CYP2D6* gene [12,13]. The characterization of these allelic variants has allowed the development of simple DNA-based tests so that individuals with a compromised capacity to metabolize CYP2D6 substrates can be identified [12,14]. Approx. 6–8% of the Caucasian population are homozygous for null alleles at the *CYP2D6* gene locus, resulting in a compromised ability to metabolize a broad range of therapeutic compounds. Such individuals have been termed 'poor metabolizers'. These studies not only demonstrate the relative importance of this enzyme in drug metabolism but also show that these genes play a central role in the disposition of foreign compounds in humans. It is therefore reasonable to hypothesize that genetic polymorphism within the cytochrome *P*-450 system will be an important factor in determining individual susceptibility to diseases for which exposure to environmental chemicals has been implicated [15].

There have been a wide range of studies to investigate the association between allelic variants of cytochrome *P*-450 *CYP2D6* and other *P*-450 genes and disease susceptibility. Allelic variants of at least six distinct human cytochrome *P*-450 genes have been identified (Table 2), and no doubt there will be many more. Following initial studies indicating an association of CYP2D6 poor metabolizers with lung cancer [16], extensive studies now indicate that this is probably not the case. However, there are some indications that this gene may be a factor in susceptibility to other cancer types, such as those of the bladder, prostate and skin. Perhaps one of the most intriguing findings with *P*-450 CYP2D6 is the potential association with susceptibility to Parkinson's disease [17,18]: poor metabolizers appear to have an approx. 2.5-fold increased risk. These epidemiological data are particularly interesting in the light of biochemical evidence indicating that *P*-450 CYP2D6 is expressed in the substantia nigra of the brain (i.e. the cells that are degraded in Parkinson's disease) and that Parkinson's-inducing chemical agents such as 1-methyl-4-phenyl-1,2,3,6-tetrahydropyridine are substrates for this enzyme [18–20]. This provides a rational association between the absence of the enzyme and an inability to detoxify such neurotoxic agents within the target cells.

In summary, there is increasing evidence demonstrating the importance of the *P*-450 system in cellular defence against environmental toxins and carcinogens. The degree to which individuality in the expression of these genes relates to individual susceptibility to environmentally linked diseases remains to be clearly established.

Table 2. Polymorphic human cytochrome *P*-450s (Caucasian population).

Enzyme	Chromosomal localization	No. of alleles	Allele frequencies
CYP1A1	15q 22-qter	4	
CYP2A6	19q 13.1-13.3	3	Consensus 17%
			v_1, T → A exon 3, 78%
			v_2, C → A exon 3, 7%
CYP2C9	10q 24.1-24.3	3	Consensus 79%
			CYP2C9*2, C → T exon 3, 12.5%
			CYP2C9*3, A → C exon 7, 8.5%
CYP2C19	10q 24.1-24.3	5	Consensus 75–79%
			$CYP2C19_{m1}$, G → A exon 5, 22–29%
			$CYP2C19_{m2}$, G → A exon 4, very rare
CYP2D6	22q 11.2-qter	> 18	Consensus 65–70%
			CYP2D6*3, del A^{2637}, approx. 4%
			CYP2D6*4, G^{1934} → A, 10–20%
			CYP2D6*5, gene depletion, approx. 4%
CYP2E1	10	3	

Glutathione S-transferases

Glutathione and glutathione-dependent enzymes play a key role in protecting cells from electrophilic chemical agents produced by the cytochrome *P*-450 system, as well as from the products of oxidative stress [6]. The glutathione transferases, similarly to the cytochrome *P*-450s, are a supergene family of proteins which inactivate toxic electrophiles through their conjugation to glutathione, or which act as peroxidases in the reduction of peroxides to water. There are now many examples where the overexpression of the glutathione S-transferases has been linked to resistance of cells to chemical damage [5,6]. Of particular interest is the association between these enzymes and the resistance of tumours to anti-cancer drugs and of normal cells to toxic and carcinogenic agents. Similarly to the cytochrome *P*-450 system, the glutathione transferases can be subdivided into gene families. At least seven gene families have now been identified in humans. Also, like the cytochrome *P*-450 system, these enzymes are polymorphic, and genetic polymorphisms have been identified at the *GSTM1*, *GSTT1* and *GSTP1* gene loci [21]. In the case of *GSTM1* and *GSTT1* there are null alleles as a consequence of gene deletions. The allele frequencies of these polymorphisms are shown in Table 3. In the case of *GSTP1*, in contrast, allelic variants involve differences in amino acid sequence. These amino acid differences have been shown to alter the substrate specificity of the enzyme towards a range of chemical agents, including the mutagenic products derived from cigarette smoke, such as benzo[*a*]pyrene diolepoxides [22]. Therefore, and similarly to the

Table 3. Polymorphic glutathione S-transferase allele frequencies in various control populations. nd, not done.

Allele	Location...	Dundee	Edinburgh	Sheffield
GSTP1				
aa		82 (37%)	75 (51%)	nd
ab		118 (53%)	62 (42%)	nd
bb		23 (10%)	10 (7%)	nd
Total		223	147	nd
GSTM1				
+		104 (43%)	nd	51 (51.5%)
Null		137 (57%)	nd	48 (48.5%)
Total		241	nd	99
GSTT1				
+		208 (86%)	112 (76%)	82 (82%)
Null		33 (14%)	35 (24%)	18 (18%)
Total		241	147	100

cytochrome *P*-450 system, genetic polymorphism in the glutathione S-transferases may be expected to be a risk factor in susceptibility to cancer.

There have now been extensive studies on this theme, and there is evidence that the *GSTM1* null allele is a risk factor in cancer of the lung, colon, skin and bladder [22]. In recent studies we have also investigated the frequency of *GSTP1* alleles in a range of human cancers and have found an increased frequency of the *GSTP1b* allele in testicular and bladder cancer [23]. In addition, an increased frequency of the *GSTP1b* allele has recently been found in lung cancer patients [24]. This increase in *GSTP1b* allele frequency also appears to be associated with increased levels of cigarette-smoke-induced carcinogen DNA adducts in lung tissue, indicating a reduced capacity of individuals carrying these alleles to inactivate carcinogens found in cigarette smoke. These epidemiological data indicated that GSTP1 may be a determinant of chemical carcinogenesis in humans.

In order to evaluate the role of GSTP1 in determining sensitivity to chemical carcinogens, we have deleted the *GSTP* genes from the mouse. In the mouse there are two *GSTP* genes (*Gstp-1* and *Gstp-2*). Both of these genes were inactivated by homologous recombination in ES cells which were used to generate *GSTP* null mice. These mice did not express any GSTP1 mRNA or any detectable GSTP proteins in any of the tissues investigated. In addition, activity towards the GSTP-specific substrate, ethacrynic acid, was absent from all tissues investigated, indicating the specificity of this particular marker substrate for this enzyme. We then carried out experiments to establish whether *GSTP1* null animals had an altered susceptibility to chemical agents. The analgesic drug paracetamol is activated by the cytochrome *P*-450 system to an imidoquinone metabolite which is hepatotoxic. This toxic product can be inactivated by conjugation with glutathione in a reaction catalysed by the glutathione S-transferases. GSTP1 has been demonstrated to be the enzyme with the highest activity towards this substrate. When a hepatotoxic dose of paracetamol was administered to the *GSTP* null animals, in contrast with the expected increase in sensitivity, the mice became resistant to its toxic effects. This appeared to be due to marked differences in the circulating and hepatic levels of the drug in the null animals. The mechanism of this effect remains to be elucidated, but the results indicate that GSTP proteins may exhibit novel functions. These data, however, demonstrate that the modulation of a single gene involved in foreign compound disposition can have a profound effect on sensitivity to chemical agents.

We have also recently studied the susceptibility of *GSTP* null mice to skin tumorigenesis induced by the polycyclic aromatic hydrocarbon dimethylbenzanthracene (DMBA). In these experiments, DMBA was administered to the skin of mice, followed by application of the tumour-promoting agent phorbol 12-myristate 13-acetate (PMA). Mice were then monitored over a 20-week period, and a profound increase in the incidence of skin papillomas was observed in the null animals (Fig. 2) [25]. Therefore GSTP can alleviate the carcinogenic effects of polycyclic aromatic hydrocarbons. These data are consistent with the epidemiological data indicating that this gene may be a factor in susceptibility to cancers of the lung and bladder induced by such agents.

Concluding remarks

In summary, a large number of genes have evolved which protect cells specifically from the potential toxic and deleterious effects of chemical agents. These protective pathways can be distinguished from the

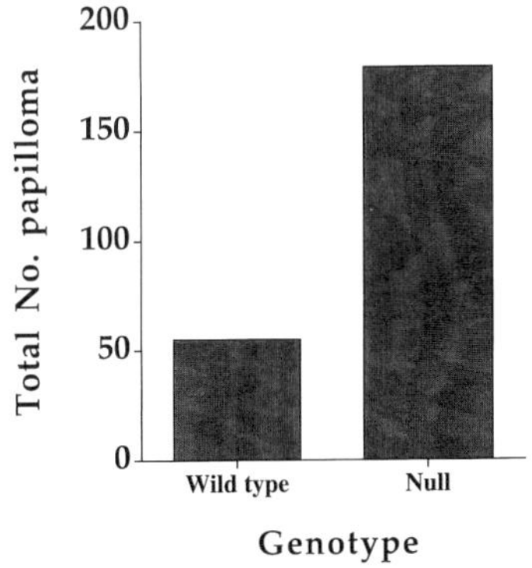

Fig. 2. Skin tumorigenesis in *GSTP1/P2* null mice. GstP1/P2(−/−) and GstP1/P2(+/+) mice were treated with a single application of DMBA (25 mg) in acteone, and thereafter twice weekly for 20 weeks with 50 μM PMA (200 ml in acetone). Papilloma numbers were scored on a weekly basis.

stress-activated pathways described in other chapters of this book. It is now clear that individual differences in the levels of expression of these genes can have a profound effect on the sensitivity of an organism to the toxic and carcinogenic effects of chemical agents. Genetic polymorphism in the expression of these genes may therefore prove to be an important factor in cancer susceptibility and in determining the sensitivity of individuals to other diseases for which exposure to environmental agents has been implicated.

G.S. thanks the Ministry of Agriculture, Fisheries and Food (Project FS1732) for financial support.

References

1. Wolf, C.R. (1986) Trends Genet. **2**, 209–214
2. Hayes, J.D. and Wolf, C.R. (1990) Biochem. J. **272**, 281–295
3. Forrester, L.M. and Wolf, C.R. (1988) in The Metabolic and Molecular Basis of Acquired Disease (Cohen, R.D., Lewis, B., Alberti, K.G.M.M. and Denman, A.M., eds.), pp. 3–18, Baillière Tindall, London
4. Keyse, S.M. (1997) in Molecular Genetics of Drug Resistance (Hayes, J.D. and Wolf, C.R., eds.), pp. 335–372, Harwood Academic Publishers, London
5. Black, S.M. and Wolf, C.R. (1991) Pharmacol. Ther. **51**, 139–154
6. Hayes, J.D. and Pulford, D.J. (1995) Crit. Rev. Biochem. Mol. Biol. **30**, 445–600
7. Ortiz de Montellano, P.R. (ed.) (1995) in Cytochrome P450 Structure, Mechanism and Biochemistry, Plenum Press, New York
8. Nelson, D.R., Koymans, L., Kamataki, T., Stegeman, J.J., Feyereisen, R., Waxman, D.J., Waterman, M.R., Gotoh, O., Coon, M.J., Estabrook, R.W., Gunsalus, I.C. and Nebert, D.W. (1996) Pharmacogenetics **6**, 1–42
9. Tucker, G.T. (1994) J. Pharm. Pharmacol. **46**, 417–424
10. Shimada, T., Yamazaki, H., Mimura, M., Inui, Y. and Guengerich, F.P. (1994) J. Pharmacol. Exp.Ther. **270**, 414–423
11. Campbell, S.J., Carlotti, F., Hall, P.A., Clark, A.J. and Wolf, C.R. (1996) J. Cell Sci. **109**, 2619–2625
12. Gough, A.C., Miles, J.S., Spurr, N.K., Moss, J.E., Gaedigk, A., Eichelbaum, M. and Wolf, C.R. (1990) Nature (London) **347**, 773–776
13. Daly, A.K., Brockmoller, J., Broly, F., Eichelbaum, M., Evans, W.E., Gonzalez, F.J., Huang, J.D., Idle, J.R., Ingelman-Sundberg, M., Ishizaki, T., et al. (1996) Pharmacogenetics **6**, 193–201
14. Spurr, N.K. and Wolf, C.R. (1992) Methods Enzymol. **206**, 149–166
15. Wolf, C.R. (1990) Cancer Surv. **9**, 437–474
16. Ayesh, R., Idle, J.R., Ritchie, J.C., Crothers, M.J. and Hetzel, M.R. (1984) Nature (London) **312**, 169–170
17. Barbeau, A., Cloutier, T., Roy, M., Plasse, L., Paris, S. and Poirier, J. (1985) Lancet **ii**, 1213–1216
18. Smith, C.A.D., Gough, A.C., Leigh, P.N., Summers, B.A., Harding, A.E., Maranganore, D.M., Sturman, S.G., Schapira, A.H.V., Williams, A.C., Spurr, N.K. and Wolf, C.R. (1992) Lancet **339**, 1375–1377
19. Gilham, D.E., Cairns, W., Paine, M.J.I., Modi, S., Poulsom, R., Roberts, G.C.K. and Wolf, C.R. (1997) Xenobiotica **27**, 111–125
20. Coleman, T., Ellis, S.W., Martin, I.J., Lennard, M.S. and Tucker, G.T. (1996) J. Pharmacol. Exp. Ther. **277**, 685–690

21. Smith, G., Stanley, L.A., Sim, E., Strange, R.C. and Wolf, C.R. (1995) Cancer Surv. **25**, 27–65
22. Hu, X., O'Donnell, R., Srivastava, S.K., Xia, H., Zimniak, P., Nanduri, B., Bleicher, R.J., Awasthi, S., Awasthi, Y.C., Ji, X. and Singh, S.V. (1997) Biochem. Biophys. Res. Commun. **235**, 424–428
23. Harries, L.W., Stubbins, M.J., Forman, D., Howard, G.C.W. and Wolf, C.R. (1997) Carcinogenesis **18**, 641–644
24. Ryberg, D., Skaug, V., Hewer, A., Phillips, D.H., Harries, L.W., Wolf, C.R., Ogreid, D., Ulvik, A., Vu, P. and Haugen, A. (1997) Carcinogenesis **18**, 1285–1289
25. Henderson, C.J., Smith, A.G., Ure, J., Brown, K., Bacon, E.J. and Wolf, C.R. (1998) Proc. Natl. Acad. Sci. U.S.A. **95**, 5275–5280

Biochem. Soc. Symp. **64**, 141–168
Printed in Great Britain

11

Cellular response to cancer chemopreventive agents: contribution of the antioxidant responsive element to the adaptive response to oxidative and chemical stress

John D. Hayes*¶, Elizabeth M. Ellis*§, Gordon E. Neal†, David J. Harrison‡ and Margaret M. Manson‡

*Biomedical Research Centre, Ninewells Hospital and Medical School, University of Dundee, Dundee DD1 9SY, Scotland, U.K., †Medical Research Council Toxicology Unit, Hodgkin Building, University of Leicester, P.O. Box 138, Leicester LE1 9HN, U.K., and ‡Sir Alastair Currie Cancer Research Campaign Laboratories, Department of Pathology, University of Edinburgh, Edinburgh EH8 9AG, Scotland, U.K.

Abstract

Cancer chemopreventive agents can act by inhibiting either the acquisition of mutations or the neoplastic processes that occur subsequent to mutagenesis. Compounds that reduce the rate at which mutations arise, referred to as blocking agents, exert their effects largely through their ability to induce the expression of antioxidant and detoxification proteins. This is achieved by the transcriptional activation of a small number of genes that are co-regulated through the presence of an antioxidant responsive element (ARE) in their promoters. Blocking agents can cause gene induction by producing oxidative and/or chemical stress within the cell

§Present address: Departments of Bioscience and Biotechnology, and Pharmaceutical Science, University of Strathclyde, Royal College, Glasgow G1 1XW, Scotland, U.K.
¶To whom correspondence should be addressed.

and, as the inducible proteins act to ameliorate the metabolic insult, the process represents a form of adaptive response. The transcription factors which mediate this response through the ARE are members of the basic leucine zipper superfamily. The mechanism whereby cells sense and respond to the chemical signal(s) generated by chemopreventive blocking agents is discussed.

Introduction: nature of cancer chemoprevention

The administration of a chemical, or dietary components, to prevent the onset or to inhibit the progression of neoplastic disease is referred to as cancer chemoprevention [1]. This type of therapy has excited considerable interest as a powerful and, in ideal situations, a relatively harmless means of providing resistance against malignant disease. Individuals who will benefit most from chemoprevention are those who have an increased risk of developing malignancy, through inherited predisposition, their age, their sex or chronic exposure to factors known to cause carcinogenesis. Examples of chemoprevention include the use of non-steroidal anti-inflammatory drugs to treat patients with adenomatous polyposis coli [2,3], and the use of the anti-oestrogen tamoxifen to treat individuals with increased risk of breast cancer [4].

A diverse number of naturally occurring and synthetic compounds possess the ability to prevent carcinogenesis. Such chemicals appear to be effective in a large number of different tissues, including liver, lung, breast, gastrointestinal tract and skin [1]. Cancer chemopreventive agents have been demonstrated to be effective at inhibiting carcinogenesis induced in rodents by polycyclic hydrocarbons, urethane, dimethylhydrazine, nitrosamines, azo dyes and aminofluorenes [5], as well as spontaneous carcinogenesis in mutant and genetically engineered mice [6,7]. The notion that minor dietary components can exert a major influence on cancer in humans is strongly supported by epidemiological studies showing that diets containing large quantities of vegetables and fruit are associated with low cancer incidence [8].

An important feature of cancer chemoprevention is that essentially all the various steps in the evolution of neoplastic disease can be inhibited by chemicals. Clearly, the use of chemical agents to inhibit the earliest stages of carcinogenesis makes this approach unique when compared with conventional cancer treatments such as surgery, radiotherapy or chemotherapy with cytotoxic drugs. Experiments carried out in laboratory animals have shown that chemopreventive agents can inhibit chemically initiated cancer at several stages of the carcinogenic process, and these agents have been divided into two broad groups depending on whether they interfere with mutagenesis or subsequent post-mutagenic steps that occur during carcinogenesis [9]. These two groups are not mutually exclusive, as certain chemicals can act at several stages of the carcinogenic process. Compounds which prevent the production of mutations are referred to as blocking agents, and include coumarins, curcumin, dithiolethiones, flavones, indoles, isothiocyanates, organosulphides, phenols, tannins and terpenes; these compounds block carcinogenesis by inducing proteins involved in detoxifying carcinogens, by inhibiting the activation of carcinogens, by preventing uptake of carcinogens, or by enhancing DNA repair. Compounds which inhibit post-mutational events in the neoplastic process are referred to as suppressing agents, and include difluoromethylornithine, isothiocyanates, carotenoids, curcumin, dehydroepiandrosterone, indoles, inositol, non-steroidal anti-inflammatory drugs, phytate, polyphenolics, retinoids, selenium, terpenes and vitamin A. These compounds have diverse actions and can act by producing differentiation, by inhibiting oncogene activation, by inhibiting the proliferation of initiated cells and by enhancing apoptosis.

The historical use of experimental animal models which involved the acute administration of chemical carcinogens to produce cancer in specific sites has led to the view that neoplasia is a multi-step process: the first step (initiation) entails a mutational event; the second step (promotion) involves selection for mutant cells that are resistant to oxidative and chemical stress; the third step (progression) involves the clonal expansion of rapidly growing resistant cells and the ultimate metastasis of such cells [10]. As all individuals possess cells carrying somatic mutations (which might be regarded as 'initiated' cells), it is often incorrectly assumed that blocking agents are likely to serve no useful function in cancer chemoprevention in humans. It is now recognized that there is no single 'initiation' event in the development of cancer. Molecular genetic studies of tumorigenesis have shown that the carcinogenic process in the human involves the accumulation of a series of genetic alterations, particularly in genes controlling cellular proliferation and the integrity of the genome [11,12], and that carcinogenesis is a continuously evolving process. Therefore many of the compounds that inhibit mutagenesis would be expected to be effective throughout the multiple stages which the development of cancer entails. It is predicted that blocking agents will prevent, or slow down, the accumulative genetic damage observed in neoplastic disease. Thus this class of agent would appear to be capable of both preventing the occurrence of, and controlling the progression of, malignant disease.

Modulation of drug metabolism and antioxidant capacity by chemopreventive agents

The majority of studies into cancer chemoprevention have been undertaken in rodents, and have focused on blocking agents rather than suppressing agents. The synthetic antioxidants butylated hydroxyanisole (BHA) and ethoxyquin have attracted a lot of attention since they were first reported by Wattenberg [13] to be excellent chemopreventive agents. The laboratory of Talalay was responsible for the discovery that, in rodents, BHA is a potent inducer of the activities of several phase II drug-metabolizing enzymes, including glutathione S-transferase (GST), epoxide hydrolase, NAD(P)H quinone oxidoreductase (NQO) and UDP-glucuronosyltransferase [14,15] (see Table 1). That this induction of drug-metabolizing enzymes accounted for the chemoprotective effect of the antioxidant was proposed by Talalay and Bueding on the basis that urine from BHA-treated animals that had been challenged with benzo[*a*]pyrene was less mutagenic in an Ames test than urine from control animals challenged with the same dose of carcinogen [14,17]. Following these early findings, it was demonstrated that many compounds with chemopreventive properties serve to induce GST and NQO activities [9].

Some of the chemopreventive agents selectively induce phase II drug-metabolizing enzymes, whereas other compounds with chemoprotective properties also induce the phase I cytochrome *P*-450 (CYP) mono-oxygenases. A further group of chemopreventive agents induce phase II drug-metabolizing enzymes while also inhibiting the activity of CYP isoenzymes. Blocking agents that only induce phase II enzymes have been called monofunctional inducers [18,19]; this group includes propyl gallate, quercetin and sinigrin [20,21] (Table 2). Agents that induce both phase I and phase II enzymes have been designated bifunctional inducers, and

Table 1. Cancer chemopreventive agents that induce phase II drug-metabolizing enzymes. The compounds listed were found to induce GST and NQO activity [16].

Class of chemical	Example	Source
Coumarins and lactones	α-Angelicalactone	*Archangelica officinalis*
	Coumarin	Leguminosae spp.
Diterpenes	Cafestol	Green coffee beans
	Kahweol	Green coffee beans
Dithiolethione	Oltipraz	Synthetic
Flavones	β-Naphthoflavone	Synthetic
	Quercetin	Citrus fruit
Indoles	Indole-3-acetonitrile	Cruciferous vegetables
	Indole-3-carbinol	Cruciferous vegetables
Isothiocyanates	Allyl isothiocyanate	Brussels sprouts
	Benzyl isothiocyanate	Garden cress
	Eugenol	Cloves, cinnamon
	Phenethyl isothiocyanate	Turnips, watercress
	Sulphoraphane	Broccoli
Organosulphides	Allyl methyl disulphide	Garlic oil
	Diallyl sulphide	Garlic oil
Phenols	BHA	Synthetic
	Butylated hydroxytoluene	Synthetic
	Ellagic acid	Grapes, strawberries
	Ethoxyquin	Synthetic
	Ferulic acid	Plums, apples, cabbages

Table 2. Chemopreventive inducers of drug-metabolizing enzymes. The data describing induction of phase I (CYP) and phase II [GST, aflatoxin B_1 aldehyde reductase (AFAR) and NQO] enzymes in rat liver are taken from [16,20,21]. It should be noted that the original definition of a bifunctional inducer was based on the ability of an agent to increase aryl hydrocarbon hydroxylase (i.e. CYP1A) activity [19]. In this Table, the term bifunctional inducer is given to compounds that induce any CYP isoenzyme, and is not restricted to members of the CYP1A family. The relative increase in the level of different enzymes is indicated by the number of plus (+) signs. A minus sign indicates a decrease in level of enzyme, and a 0 indicates no significant change. Abbreviation: nd, not determined.

Chemopreventive agent	Enzyme induction in rat liver (relative to control)						
	CYP1A	CYP2B	CYP3A	CYP4A	GST	AFAR	NQO
Monofunctional inducers							
Propyl gallate	0	nd	nd	nd	+	0	+
Quercetin	0	nd	nd	nd	+	+	0
Sinigrin	0	nd	nd	nd	+	++	+
Bifunctional inducers							
Benzyl isothiocyanate	—	0	+	++	++	+++	nd
Butylated hydroxytoluene	0	+++++	+++	—	+++	+++	++++++
Caffeic acid	+	0	+	++	+	—	0
Coumarin	—	0	+	0	+++++	+++++	nd
Diallyl sulphide	—	+++	+	—	+	+	nd
Ethoxyquin	++	+++++	+++	+	++++	++++	+++
Indole-3-carbinol	+++++	+++++	++	0	+++	++	+++++
Oltipraz	+	++	+	—	+++	+++	++++
Phenethyl isothiocyanate	+	+++	0	0	+	++	+++
Dual-action agents							
Allyl disulphide	—	0	0	—	+	++	nd
BHA	—	0	0	0	+++	+++	++++++
4-Methoxyphenol	—	—	—	—	+	++	nd
4-Methyl catechol	—	nd	nd	nd	++	+	0
α-Tocopherol	—	nd	nd	nd	++	+	+

include ethoxyquin, indole-3-carbinol, isothiocyanates and oltipraz. Compounds that induce phase II enzymes while inhibiting phase I enzymes have been called dual-acting agents [21], and this group includes BHA, 4-methylcatechol and α-tocopherol (Table 2). As bifunctional inducers can increase CYP isoenzymes associated with the activation of carcinogens, it has been suggested that their beneficial effects may be outweighed by their potential deleterious effects. This possible concern is probably overstated, as it assumes that the chemicals activated by CYP isoenzymes represent the major cause of carcinogenesis in humans, an assumption that is not supported by epidemiological data. Bifunctional inducers can be highly effective chemopreventive agents. The effectiveness of cancer chemopreventive blocking agents appears to be determined primarily by their ability to increase phase II conjugating enzymes to a greater extent than phase I mono-oxygenases. For example, indole-3-carbinol is a bifunctional inducer, but has been shown to be an excellent chemopreventive agent [21].

The observation that many chemopreventive agents induce drug-metabolizing enzymes has led to the assumption that the benefits that they confer result solely from enhancement of the ability of the cell to protect itself against noxious foreign compounds. However, possibly more important than enhanced metabolism of xenobiotics is the fact that many of the inducible phase II drug-metabolizing enzymes exhibit antioxidant functions and will protect the cell not only against reactive oxygen species (ROS), such as superoxide anion, H_2O_2 and the hydroxyl radical, but also against toxic by-products generated by free radicals interacting with macromolecules. For example, certain inducible GSTs exhibit glutathione peroxidase activity and detoxify α,β-unsaturated carbonyls and epoxides produced by oxidative stress [16]. The inducible NQO also has antioxidant properties through its ability to form redox-stable hydroquinones [22]; however, this enzyme can also form redox-labile hydroquinones which need to be detoxified by other phase II enzymes such as GSTs. Chemopreventive agents can increase significantly the intracellular level of GSH, the major non-protein thiol with strong antioxidant properties [23]. The enzyme γ-glutamylcysteine synthase (GCS), which catalyses the rate-limiting step in the synthesis of glutathione, is inducible by BHA [24]. The chemopreventive agent 1,2-dithiole-3-thione increases the antioxidant capacity of rat liver by inducing haem oxygenase-1 (HO-1) as well as the ferritin L and H chains [25]; the combined increases in HO-1 and ferritin remove the pro-oxidant haem and sequester the haem iron, thereby preventing the generation of free radicals through the Fenton reaction [26]. These data suggest that an important aspect of cancer chemoprevention is protection against oxidative stress.

It is important to recognize that inducers of phase II enzymes include many non-nutrient phytochemicals (Table 1) that are present in fruit and vegetables, which epidemiological studies have associated with reduced cancer risk [8]. This indicates that the induction of drug-metabolizing and antioxidant enzymes by phytochemicals represents at least one of the mechanisms whereby fruit and vegetables can protect against carcinogenesis.

Identification of genes responsive to chemopreventive agents

Most phase II drug-metabolizing enzyme systems represent multigene families, and few studies have been undertaken to identify specific genes involved in resistance to particular carcinogens. We have investigated mechanisms of chemoprevention against the hepatocarcinogen aflatoxin B_1, a naturally occurring mycotoxin which is widely distributed in humid areas of the world such as Asia and Africa, and which along with hepatitis B virus is thought to be responsible for the high incidence of liver cancer in these regions [27]. Rats fed on normal laboratory chow are highly sensitive to aflatoxin and will readily develop liver cancer within 12 months if exposed to as little as 4 p.p.m. aflatoxin B_1 in the diet for a 6-week period. However, inclusion of the antioxidant ethoxyquin (0.5%) in the diet during exposure to aflatoxin B_1 prevents the formation of preneoplastic foci and liver tumours [28]. As the failure of aflatoxin to produce hepatomas was found to be associated with both a marked decrease in its ability to form DNA adducts and an increase in the amount of aflatoxin-B_1–glutathione conjugate secreted into bile, it was proposed that ethoxyquin conferred protection by altering the primary metabolism of aflatoxin B_1 and inducing phase II enzymes, including GST [29,30]. The ethoxyquin-inducible transferases in rat liver that are responsible for catalysing this conjugation are GSTA2-5 and GSTA3-5 [31]. As these two dimeric enzymes share the A5 subunit, it was suggested that this polypeptide is responsible for detoxifying aflatoxin B_1-8,9-epoxide, the ultimate carcinogen. This was confirmed by molecular cloning and heterologous expression of GSTA5 [32]. During study of the biochemical basis of chemoprevention by ethoxyquin, a novel inducible aflatoxin aldehyde reductase (designated AFAR) was characterized which catalyses the reduction of the dialdehydic form of aflatoxin B_1-8,9-dihydrodiol. This dialdehyde is thought to be harmful to the liver because it can form Schiff bases with lysine residues in protein [33,34]. It is now proposed that ethoxyquin confers protection against the genotoxic effects of aflatoxin B_1 by induction of GSTA5 and against its cytotoxic effects by induction of AFAR [33].

Among the cytosolic detoxification enzymes, GST and NQO are highly responsive to chemopreventive agents. The properties of the inducible GST isoenzymes have been reviewed relatively recently [16]. Two NQO isoenzymes have been described, designated NQO1 and NQO2, and of these the former represents the inducible protein ([35,36]; for a review of the early literature, see [37]). In the rat, AFAR, GSTA2, GSTM1, NQO1 and microsomal epoxide hydrolase are typically the phase II enzymes that are most inducible by the compounds listed in Table 1 [16,37]. Rat GSTP1 is also highly induced by synthetic antioxidants, dithiolethiones, diterpenes and polychlorinated biphenyls [16,38,39]. In the mouse, GSTA1, GSTA2, GSTM1, GSTM2, GSTM3, GSTP1/P2 and NQO1 have been found to be highly inducible by foreign and naturally occurring chemicals [40–42].

Induction of phase II detoxification enzymes by chemopreventive agents is not unique to rodents, although studies in humans are limited because of difficulty in sampling tissue. O'Dwyer et al. [43] demonstrated significant increases in NQO1 and GCS in human colon mucosal biopsies from patients at risk of colorectal cancer who were treated with oltipraz. The amount of GSTP1 has also been shown to be modestly increased in human colon mucosal biopsies from healthy volunteers who ate diets enriched with Brussels sprouts [44]. Tew and colleagues have identified a highly inducible member of the aldo–keto reductase family, 20α-hydroxysteroid dehydrogenase, in human colonic HT-29 cells [45]. In primary human hepatocytes, marked induction of GSTA1/A2 mRNA by 1,2-dithiole-3-thione and sulphoraphane has been observed [46,47], and in Hep-G2 cells the GSTA1 and GSTA2 subunits are both inducible by picolines (methylated pyridines) [48].

Characterization of the antioxidant responsive element (ARE) as an element that mediates responsiveness to chemopreventive agents

Induction of drug-metabolizing and antioxidant enzymes by compounds classed as chemopreventive blocking agents is frequently mediated by the ARE. The term ARE is possibly too restrictive, as it has become apparent that this element responds to many classes of chemicals, including antioxidants, oxidants and electrophiles. This element was first recognized by Pickett and his colleagues as a region of DNA responsible for both basal and inducible expression of the rat *GSTA2* gene [49,50]. The ARE was found to be located within a 41 bp region flanking rat *GSTA2* (located between −722 bp and −682 bp from the transcriptional start site) that was originally shown to be responsible for part of the induction of the gene by β-naphthoflavone [50]. Rushmore and Pickett [51] demonstrated that the ARE is also responsive to t-butylhydroquinone (tBHQ), a major metabolite of BHA. Subsequent deletion analysis of the 5′-flanking region of *GSTA2* showed that responsiveness to β-naphthoflavone and tBHQ resides in a 10 bp region (5′-GTGACAAAGC-3′), and mutation analysis defined the core enhancer sequence to be 5′-(G/A)TGACNNNGC-3′ [52] (Fig. 1). It is important to note that, while the ARE affects both basal and inducible expression, these two functions are not superimposable. Thus TGAC (nucleotides −696 to −693 of *GSTA2*) is essential for both basal and inducible expression, whereas GC (nucleotides −689 and −688 of *GSTA2*) is essential for induction but not basal expression [52].

Accumulating data indicate that the function of the core ARE is influenced by adjacent 5′ and 3′ sequences. Figure 1 shows that, 6 bp upstream from the ARE, a 10 bp sequence can be identified which is closely similar to the core enhancer. This related sequence, which is tandemly arranged with the core ARE, appears to enhance basal expression, but is not required for induction [52,57]. In addition to the upstream ARE-related sequence, an AT-rich region can be identified immediately downstream from the core enhancer (Fig. 1). The precise function of this AT box requires clarification, but the number of adenine nucleotides within the box correlates positively with the level of basal expression and appears to have a minimal effect on inducible expression [58].

The ARE flanking rat *GSTA2* was first described at a conference in Edinburgh in 1989 [49]. The following year, a report from the laboratory of Daniel identified a closely related sequence flanking the murine *GstA2* gene that they designated the electrophile responsive element [53]. Within the 41 bp region two differences exist between the murine and rat sequences, one of which results in the formation of a second upstream perfect ARE (see Fig. 1). More recent work by Wasserman and Fahl [57] suggests that the two ARE sequences flanking *GstA2* are not functionally

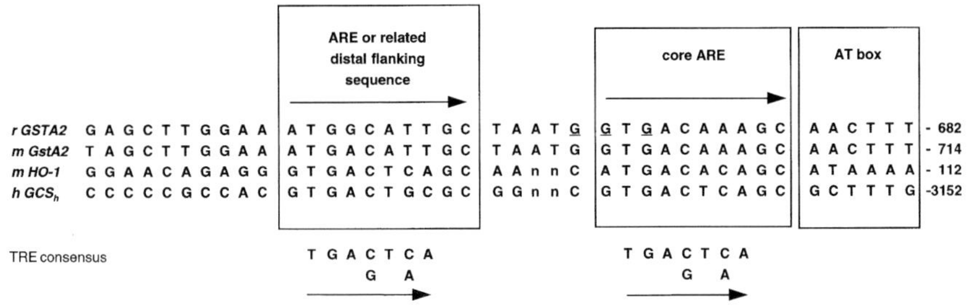

Fig. 1. AREs: enhancer found in tandem repeat orientation. Enhancers in the 5′-flanking region of four genes that are inducible by chemopreventive agents are shown aligned with the original 41 bp element described by Rushmore et al. [50]. The data for rat *GSTA2*, mouse *GstA2*, mouse *HO-1* and human *GCS$_h$* are from Rushmore et al. [52], Friling et al. [53], Prestera et al. [54] and Mulcahy et al. [55] respectively. The core ARE, the distal ARE-like motif and the proximal AT flanking sequence are boxed separately. The TRE consensus is aligned below to allow comparison with the ARE. The nucleotides in the two *GST* genes are numbered from the transcriptional start site, and are shown in the correct orientation. The numbering of nucleotides in *HO-1* is from the origin of replication of pMHOSX2cat. The 5′-flanking region of *GCS$_h$* is shown in reverse orientation and is numbered from −3083 to −3152. In the regions of *HO-1* and *GCS$_h$* between the core ARE and the ARE-like sequence, 39 and 31 nucleotides respectively (represented by nn) have been omitted. The nucleotides in rat *GSTA2* that are underlined have been shown by methylation interference and protection assays to be involved in high-affinity binding by nuclear extracts from Hep-G2 cells [56].

identical, as transcriptional activation by tBHQ is achieved by the proximal ARE, whereas the distal upstream ARE appears to serve an amplifying role. These workers confirmed that the distal ARE enhances transcriptional activation of mouse *GstA2* by tBHQ and that this property resides in the sequence 5′-ATGACATT-3′ contained within the upstream ARE. The *in vivo* consequence of differences in the number of core ARE sequences in gene promoters has not been established, but by comparison with the rat GSTA2 subunit the mouse GSTA2 subunit shows relatively greater induction by BHA.

The identity of the nucleotide in the AT box that is immediately adjacent to the core ARE may attenuate induction. In their original mutational analysis of the rat *GSTA2* ARE, Rushmore et al. [52] showed that neither basal expression nor induction by β-naphthoflavone was affected by the presence of an A or a G at position −687; thus this nucleotide was not included in the consensus sequence. More recently it has been shown that mutation of A to C at the equivalent position in mouse *GstA2* (i.e. −719) reduces the level of reporter gene expression following tBHQ exposure to 10% [57]. Unfortunately, it is not clear whether this A to C mutation at −719 in *GstA2* affects basal expression or inducible expression. Studies by Favreau and Pickett [58] on the AT box in the rat *NQO1* gene implied that mutation of A to C would reduce basal expression and would not alter induction. Further studies of the functional significance of this nucleotide have not been reported, and it has not been established whether mutation of the nucleotide immediately downstream from the core ARE from an A to a T still allows activation by tBHQ and other inducers. It should be noted that in GCS_h (the gene encoding the GCS heavy subunit) a G exists as the nucleotide in the AT box immediately adjacent to the core ARE (Fig. 1). These data suggest that the ARE is possibly more accurately defined as 5′-(G/A)TGACNN-NGC(G/A)-3′.

Following the characterization of the *cis*-acting elements that are responsible for the induction of rat and mouse GST genes by phenolic antioxidants, analyses of the promoters of the murine *HO-1* gene and GCS_h have revealed the presence of functionally active ARE sequences that are also tandemly arrayed [54,55]. By contrast with the core ARE and related sequences flanking *GSTA2* and *GstA2*, which are separated by only 5 bp, those flanking the mouse *HO-1* and human GCS_h genes are situated almost 40 bp apart. The functional significance of this difference is unclear.

In addition to enhancers comprising tandemly arrayed ARE sequences, a number of gene promoters exist in which the core ARE is situated immediately downstream from an inverted ARE half-site (Fig. 2). Examination of the flanking region of the rat and human *NQO1* genes shows that this inverted ARE forms a 13 bp palindromic sequence 5′-AGTCACAGTGACT-3′, where the last 6 bp are part of the core ARE [59,60]. Mutational analysis has revealed that the distal ARE half-site is essential for induction by β-naphthoflavone, and loss of this half-site possibly has a greater effect on inducibility than does loss of the distal

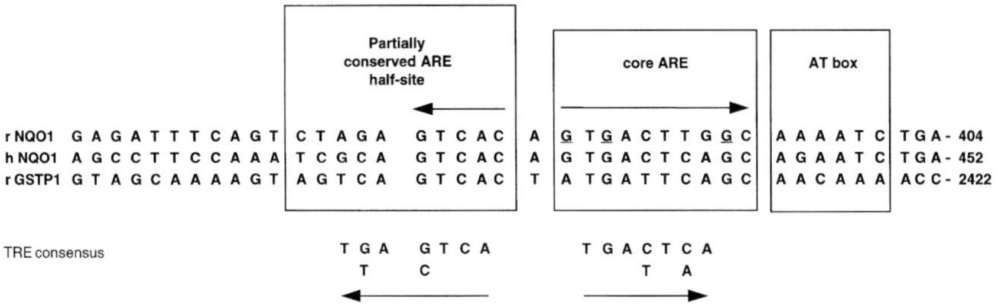

Fig. 2. AREs: enhancer found in inverted repeat orientation. Alignment of enhancers from three genes that are responsive to chemoprotective agents. Data for rat *NQO1*, human *NQO1* and rat *GSTP1* are from Favreau and Pickett [59], Jaiswal [60] and Okuda et al. [61]. The four conserved nucleotides in the distal 'partially conserved ARE' are grouped separately and their inverse orientation is depicted by an arrow. To allow comparison between the ARE and TRE, the consensus TRE is shown at the bottom of the figure, and its orientation is indicated by an arrow. The maximum distance between the distal ARE half-site and the proximal core ARE has been demonstrated in *GSTP1* to be 7 bp [62]. Nucleotides in rat *NQO1* that are underlined have been shown by methylation interference and protection assays to be required for high-affinity binding by nuclear protein from Hep-G2 cells [63].

ARE-related sequence in the tandemly arranged enhancers [58]. Furthermore, the first adenine in the AT box (see Fig. 2) is essential for inducibility, and deletion of other adenine nucleotides in this box results in the progressive loss of basal gene expression. At present it is not clear whether it is mechanistically helpful to distinguish between palindromic and tandemly arrayed AREs. However, Favreau and Pickett [58] have shown that in Hep-G2 cells the palindromic ARE is a more potent enhancer than is the tandemly arrayed ARE, an effect that cannot be entirely attributed to differences in the AT box.

The rat *GSTP1* gene has been studied because the subunit it encodes provides an early marker for hepatocellular carcinoma [16]. The transcriptional activation of this gene during neoplasia was attributed by Muramatsu and his co-workers to an element referred to as GPEI [61], a hypothesis subsequently confirmed using several distinct lines of transgenic rats [64]. GPEI does not conform to a perfect ARE as originally defined by Rushmore et al. [52]. However, it is closely related to the palindromic enhancers found flanking the rat and human *NQO1* genes (Fig. 2). Favreau and Pickett [58] have found that GPEI is responsive to tBHQ and have shown that the factors that bind the ARE can be competed for by GPEI-containing oligonucleotides in electrophoretic mobility shift assays (EMSAs). Together, these data suggest that the consensus ARE should be defined as 5′-(G/A)TGA(C/T)NNNGC(G/A)-3′. The observation that GPEI can function as an ARE is important, as it would indicate that this enhancer may be able to act as a site of integration of several signals, including those associated with preneoplasia.

Transcription factors involved in ARE function

In view of the similarity between the ARE and the TPA-responsive element [TRE; where TPA is another name for PMA (phorbol 12-myristate 13-acetate)] [5′-TGA(C/G)T(A/C)A-3′] [65], a number of investigators have examined the possibility that activator protein-1 (AP-1) may be involved in transcriptional activation of genes by the ARE. Several laboratories have concluded that Jun–Fos heterodimers do not bind the core ARE [66–68]. However, evidence exists that other members of the basic region leucine zipper (bZIP) superfamily are able to bind the ARE. The nuclear-factor-E2-related factors Nrf1 and Nrf2, which are related to the *Drosophila* cap and collar (CNC) factor [69], have been shown by transfection experiments to be capable of mediating induction by β-naphthoflavone and tBHQ in Hep-G2 cells [70]. Further, the nuclear factor-E2 binding site contains an ARE [71,72]. Compelling evidence that these transcription factors are crucial for enzyme induction *in vivo* has come from the work of Itoh et al. [73], who generated an *Nrf2* gene knockout mouse and showed that BHA was ineffective at inducing GST in mice homozygous null for *Nrf2*.

Nrf2 does not bind the ARE as a homodimer, but rather is required to dimerize with other bZIP factors. Itoh et al. [73] have described EMSA experiments, using purified heterologously expressed transcription factors, showing that it is necessary for Nrf2 to dimerize with a small Maf protein in order to bind the ARE. In these studies it was shown that mouse Nrf2 can form a heterodimer with MafK which can 'band-shift' the ARE in mouse *GstA2*. It has been proposed that both human Nrf1 and Nrf2 can dimerize with either MafG or MafK [74], although *in vivo* the formation of heterodimers will depend on the tissue-specific expression of the factors. Also, MafG is itself inducible by H_2O_2, and presumably the composition of the heterodimer will be influenced by the extent of oxidative stress in the cell [75]. Further research is required to establish whether Nrf1 interacts with the ARE *in vivo* and which of the small Maf proteins are involved in enzyme induction.

Although evidence indicates that transcriptional activation of *GST* and *NQO* gene expression through the ARE is mediated by Nrf and small Maf proteins, it is clear that other factors can interact with this element, at least *in vitro*. Photochemical cross-linking of nuclear proteins to the ARE showed that the enhancer was bound *in vitro* by polypeptides of 45 kDa and 28 kDa [56], which are smaller than Nrf1 and Nrf2 [70] but larger than MafG and MafK [76]. The reason for the differences in estimates of the size of the ARE-binding proteins is not known, but it is possible that other members of the CNC and small Maf protein families can bind this element. In addition to Nrf1 and Nrf2, p45 (which can also dimerize with small Maf proteins [72]) may bind the ARE. Combinations of gel-filtration chromatography and EMSA have revealed that protein complexes of 74 kDa and 160 kDa can bind the ARE [77–79]. Wasserman and Fahl [57] have produced a model in which bZIP factors bind the core ARE but separate proteins interact with adjacent flanking sequences. If this model is correct, it might be expected that tandemly arranged AREs would recruit different factors from inverted repeat AREs. Indeed, results from DNA methylation interference and protection data suggest that the binding of nuclear proteins to the tandemly arrayed rat *GSTA2* ARE (interaction with 5′-GGTGACAAAGCA-3′ [56]) is distinct from that which occurs with the palindromic ARE flanking the rat *NQO1* gene (interaction with 5′-AGTGACTTGGCA-3′ [63]). It remains to be established whether multiple bZIP proteins interact with the ARE. However, heterogeneity in ARE-binding proteins is apparent from the fact that the high-affinity protein–ARE complex obtained from crude nuclear extracts from mouse hepatoma Hepa 1c1c7 cells was found to be electrophoretically distinct from that observed using extracts from mouse F9 cells [58].

It is important to recognize that the core ARE can co-exist with a TRE, and in the case of the human *NQO1* gene both elements are found within the same 24 bp of the promoter. In this situation, binding by multiple *trans*-acting factors to the ARE can occur *in vitro*, which complicates the identification of complexes associated with induction by chemopreventive agents. Jaiswal and colleagues have used EMSAs to show that antibodies against Jun and Fos will supershift complexes binding the bipartite ARE/TRE enhancer found in human *NQO1* [70]. In the case of Fos, these workers have found that it negatively regulates the expression of *NQO1*, but it is not known whether the effect is direct or indirect.

Purification of factors that bind the rat *GSTP1* enhancer GPEI and its closely associated upstream region has shown that c-Jun and a novel 55 kDa protein, which is immunochemically related to the Maf family, can interact *in vitro* with GPEI [80]. Whether these proteins mediate the induction of *GSTP1* by antioxidants remains to be established.

The ARE responds to several groups of chemopreventive agents, most of which are thiol-active

Functional examination of the ARE using reporter gene constructs has demonstrated that it is responsive to a remarkably large number of different compounds, the diversity of which excludes the existence of a simple receptor-based mechanism. The chemicals that transcriptionally activate gene expression via the ARE can be divided into two major groups, namely direct-acting and indirect-acting inducers, depending on whether they need to be metabolized in order to be effective [16].

The direct-acting ARE inducers, although probably comprising several hundred compounds, can be grouped into five classes on the basis of common chemical features (Fig. 3). The research groups of Talalay and Pickett pointed out that these inducers are redox cycling agents (e.g. quinones and catechols) [51], pro-oxidants (H_2O_2, hydroperoxides, heavy-metal salts and arsenicals) [51,81], soft electrophiles (e.g. isothiocyanates and Michael reaction acceptors) [82], mercaptans (principally dithiols) [81] or organosulphurs (dithiolethiones) [81].

Certain chemopreventive agents have been proposed to induce oxidative stress by increasing the production of ROS such as superoxide anion, H_2O_2 or the hydroxyl radical. EPR spin trapping has demonstrated that tBHQ can produce free radicals in Hep-G2 cells, and the finding that inclusion of catalase in the culture medium inhibits activation of the mouse *GstA2* ARE by tBHQ in these cells suggests that the generation of ROS (i.e. H_2O_2) is responsible for induction [83]; it is believed that catalase exerts its effects outside the cell, and that it reduces the level of H_2O_2 that diffuses across the membrane. It should, however, be stressed

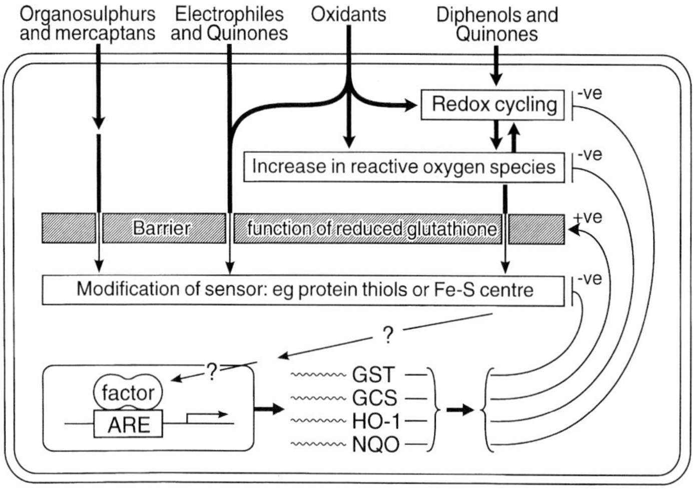

Fig. 3. Changes in intracellular metabolism that are associated with exposure to chemopreventive enzyme inducers. The various classes of inducing agent that affect transcriptional activation via the ARE can influence the thiol status of the cell in a variety of ways, some of which are represented by open boxes. The vertical line drawn between organosulphurs and the sensor is broken because it is uncertain whether such compounds act directly or need to be metabolized in order to be effective. Reduced glutathione antagonizes ARE-driven gene expression [83,84] and, as it is presumed to serve a barrier function, it is represented by a shaded box. It is not clear how alterations in thiol status are signalled to the ARE, but it is postulated that the presence of inducing agents is detected by the cell through a nucleophilic centre, possibly a sensitive cysteine residue(s) in one or more sensor proteins. The location of the sensor protein(s) is not known, and the diagram is not intended to imply that it is necessarily a cytosolic protein. Collectively, the genes that are activated through the ARE function to correct metabolic imbalances, such as GSH depletion and increases in ROS, caused by the inducing agent. Thus the co-ordinated induction of GST, GCS, HO-1 and NQO represents a form of adaptive response, although each enzyme acts in a distinct fashion to correct different forms of stress.

that generation of ROS is not a universal property of inducers, as it is doubtful whether isothiocyanates or mercaptans will generate free radicals.

A general feature of essentially all direct-acting ARE inducers is their ability to interact with thiols. Although most of these inducers may be capable of altering GSH levels, the fact that this intracellular thiol is abundant argues against GSH depletion representing the primary stimulus for gene activation via the ARE. Specifically, in the presence of millimolar concentrations of GSH within the cell, it is difficult to imagine that compounds such as benzyl isothiocyanate at concentrations of less than 10 μmol/l can activate gene expression by lowering the level of glutathione [81]. Nonetheless, the importance of intracellular thiol levels in this response cannot be ignored, as depletion of GSH in Hep-G2 cells by treatment with buthionine-*S*,*R*-sulphoximine or the oxidizing agent diamide stimulates transcription driven by the mouse ARE [83]. Furthermore, *N*-acetylcysteine treatment attenuates ARE-mediated transcriptional activation by hydroquinone and dimethylfumarate, compounds that are considered to be direct-acting inducers [84]. The observation that GSH levels are negatively correlated with ARE-driven gene expression suggests that this thiol serves a barrier function and antagonizes ARE-driven transcription. Since GSH depletion does not appear to be the primary effector of enzyme induction, it is possible that a separate critical thiol, or thiols, exists in some compartment of the cell which acts as a sensor to monitor the level of intracellular oxidative and chemical stress.

Indirect-acting ARE inducers need to be oxidized by CYPs, or other enzymes, in order to effect induction by the ARE (Fig. 4). For example, metabolism of β-naphthoflavone involves oxidation by the inducible CYP1A1, a reaction that requires prior induction of the oxidase by the substrate through ligand binding to the aryl-hydrocarbon (Ah) receptor followed by interaction with the xenobiotic responsive element flanking the *CYP* gene [51]. At present, the nature of the chemical signal produced by oxidation of these CYP substrates is uncertain, and it is not clear whether induction is achieved by a specific metabolite or by the generation of free radicals arising from poor coupling between CYP and the reductase.

The conversion of an indirect-acting ARE inducer into an active inducer is not a unique function of CYP1A1, and other indirect-acting ARE inducers such as BHA, ethoxyquin and coumarin are metabolized to active inducers by different members of the CYP superfamily. Although the identity of the CYP isoenzymes responsible for converting BHA into an ARE inducer is not proven, evidence suggests that its O-de-methylation to form tBHQ is essential for induction. In Hep-G2 cells, tBHQ through redox cycling to t-butylquinone is able to generate the hydroxyl radical, whereas BHA cannot carry out this reaction [84]. In this discussion it is important to note that tBHQ also forms glutathione conjugates [85] and therefore, in addition to·redox cycling, it may also modify intracellular thiols. Circumstantial evidence suggests that the antioxidant ethoxyquin must be O-de-ethylated to form 6 hydroxy

2,2,4-trimethyl-1,2-dihydroquinoline and subsequently converted into 2,2,4-trimethyl-6-quinolone for induction to occur [86]. It is highly likely that coumarin, which is a potent inducer of GST and AFAR in the rat, also needs to be metabolized by CYPs in order to be effective, but in this case the metabolic fate of coumarin is subject to marked species variation (coumarin is inactive in Hep-G2 cells [81]); in humans 7-hydroxylation is a major pathway, whereas in the rat 3-hydroxylation is a principal route of metabolism [87].

The anti-oestrogen tamoxifen can activate ARE-driven gene expression in breast cancer cells [88]. This process is dependent on the presence of the oestrogen receptor, and can therefore be distinguished from activa-

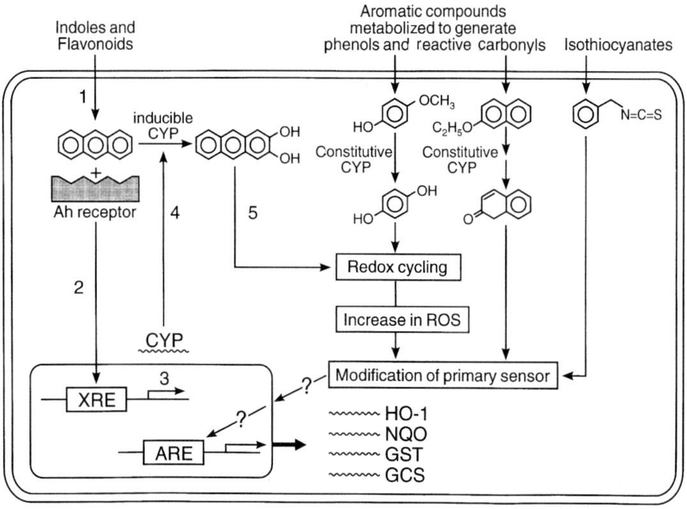

Fig. 4. Role of CYP isoenzymes in the activation of chemo-preventive enzyme inducers. The structures of the chemicals shown are representative of the types of inducers involved in chemoprevention, and are not intended to represent specific compounds. Inducers that transcriptionally activate genes through the ARE may function directly (e.g. isothiocyanates and mercaptans). Alternatively they may require metabolic activation, and are either converted into active inducing agents by CYP isoenzymes that are normally constitutively expressed (e.g. BHA and ethoxyquin in rat liver), or converted into the active inducing agent by CYP isoenzymes that are not normally expressed and therefore require prior induction themselves (e.g. flavonoids and indoles). Chemopreventive agents that need to be oxidized by inducible CYP1A enzymes before they can activate gene expression through the ARE trigger the following steps in order to be effective: (1) binding of chemopreventive agent to the aryl-hydrocarbon (Ah) receptor; (2) translocation of the Ah-receptor–chemo-preventive-agent complex to the nucleus; (3) transcriptional activation of xenobiotic responsive element (XRE)-regulated *CYP1A* genes; (4) oxidation of the agent by CYP1A; (5) stimulation of chemical-stress signalling pathway by metabolites generated by CYP1A to effect induction via the ARE.

tion by typical inducers that work via the ARE. It is not known whether the ability of tamoxifen to act as an inducer of phase II drug-metabolizing enzymes is a direct or indirect effect. The fact that tamoxifen requires the oestrogen receptor in order to transcriptionally activate genes through the ARE suggests that the receptor itself may interact with Nrf/Maf, or that co-activators, such as CBP or p300, may be involved [89]. Certainly, cross-talk can occur between nuclear hormone receptors and the nuclear factor-E2 bZIP protein [90].

ARE-mediated induction of antioxidant and detoxification enzymes represents an adaptive response

The fact that a number of enzymes which are regulated through the ARE have antioxidant and detoxification functions, and that the majority of the compounds which act through the ARE are pro-oxidants and/or electrophiles, suggests that induction is an adaptive response to eliminate oxidative or chemical stress. Induction of GST, NQO, HO-1 and GCS probably represents a 'feedback loop' to enable the cell to correct the redox imbalance or the accumulation of thiol-modifying compounds resulting from exposure to inducers. This hypothesis is supported by the work of Primiano et al. [25], who showed that 1,2-dithiole-3-thione, a compound demonstrated to produce transcriptional activation through the ARE, induces not only detoxification genes but also antioxidant genes, including ferritin L and H chains and HO-1.

Sensor and signalling pathways involved in ARE-mediated induction of gene expression

Having established that the transcriptional response of cells to cancer chemopreventive inducing agents is probably mediated by an Nrf/small-Maf heterodimer through the ARE, four important overlapping questions remain to be addressed. Firstly, do all ARE-specific chemopreventive inducing agents act by producing a single unique form of oxidative/chemical stress? Secondly, how do cells sense the presence of chemopreventive agents? Thirdly, once a response is triggered, how is the signal relayed to the transcriptional machinery? Fourthly, do all cancer chemopreventive agents use the same signalling pathway?

On the basis of their chemical properties, most chemopreventive inducing agents can be defined in two main groups, namely pro-oxidants (redox-cycling drugs, metabolizable polyphenolics, peroxides [52]) and soft electrophiles (isothiocyanates, Michael reaction acceptors [82]). As both groups of agents contain thiol-active compounds, it has been proposed that they all produce a form of oxidative/chemical stress, and that this stress is sensed by the cell possibly through a nucleophilic centre in a cellular macromolecule, probably a protein.

It is possible that Nrf/small-Maf heterodimers act themselves as the primary sensor of the types of inducers described above, thereby obviating the requirement for a separate detector and signalling process. However, in mammalian cells there are no examples of transcription factors interacting directly with oxidants and electrophiles in a fashion that would allow induction of gene expression. It is more likely that Nrf/Maf heterodimers are activated indirectly by a protein kinase cascade or a redox cascade, or both. Evidence has been presented that transcriptional activation of gene expression by β-naphthoflavone and tBHQ through the ARE is indirect and may require a signalling pathway that involves protein kinases. Specifically, treatment of Hep-G2 cells with the phosphatase inhibitor okadaic acid can mimic induction by β-naphthoflavone and tBHQ, while the protein tyrosine inhibitor genistein is able to abolish induction [91]. Examination of the primary structures of Nrf1 [92] and Nrf2 [93] reveals the presence of a number of potential phosphorylation sites within their *trans*-activation domains, indicating that chemopreventive agents possibly induce gene expression through the ARE by activating signal transduction pathways which result in the modification of these factors. By contrast,

(a)

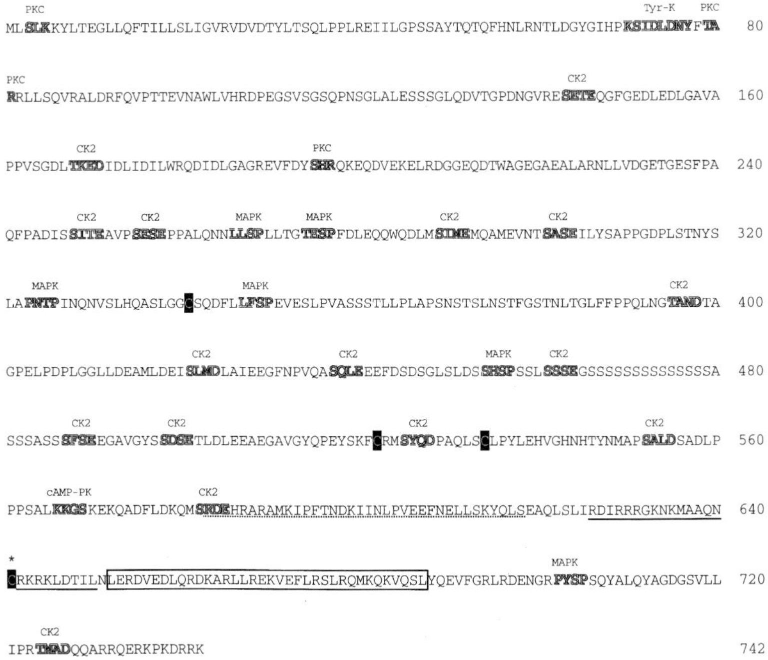

Fig. 5. Primary structures of human Nrf1 (a) and Nrf2 (b) showing potential phosphorylation and redox sites. The primary structures of Nrf1 and Nrf2 have been deduced from sequencing their cDNAs [92,93]. The putative *trans*-activation domains lie in the N-terminal portion of the proteins, which are rich in glutamic acid and aspartic acid residues. Residues that represent the 'cap and collar' homology region are shown as dotted

small Maf proteins lack *trans*-activation domains, but the presence of a serine/threonine-rich region at the C-terminus [94] suggests that these may also be phosphorylated. Both Nrf and Maf contain a number of cysteine residues, one of which is conserved and has been implicated in the redox sensitivity of DNA binding by other bZIP proteins; these cysteine residues may be part of a redox signalling cascade that results in activation of the factors. Figure 5 shows the potential phosphorylation sites and redox-sensitive cysteines in Nrf1 and Nrf2.

If the ARE-binding transcription factors do not respond directly to the chemopreventive inducing agent, then the role of primary sensor of the cell that triggers a response could be performed by one or more proteins that contain reactive cysteine residues that are modified by the oxidative/chemical stress produced by the inducing agent. It is also possible that this function could be carried out by a protein containing a redox-sensitive iron–sulphur cluster, as described for the bacterial SoxR sensor/regulator [95,96]. While the putative sensor may be a highly specialized protein, it is more likely that this function is performed by protein(s) whose function is not uniquely associated with enzyme induc-

(b)

Fig. 5. (*continued*)
underlined, residues that comprise the basic DNA-binding domain are underscored with a solid line, and the residues that form the leucine zipper are boxed. Potential sites of phosphorylation by cAMP-dependent protein kinase (cAMP-PK), casein kinase II (CK2), MAPK, protein kinase C (PKC) and tyrosine kinase (Tyr-K) are indicated. The cysteine residues are shown as white on a black background, and that in the basic region is indicated by an asterisk.

tion, but is part of a more general signalling pathway. Indeed, there are some interesting candidates for the role of sensor among protein kinases, protein phosphatases, cell surface receptors and reductases, whose activities are known to be sensitive to chemical modification by oxidants or electrophiles. An alternative possibility is that the sensor may respond to glutathione conjugates rather than oxidants/electrophiles, but this would suggest a more specialist role for the protein. It is apparent that an important issue that remains to be settled is whether the cell contains multiple types of sensor, possibly with different specificities.

Oxidative stress and alkylating agents have been shown to activate several receptor tyrosine kinases situated at the inner surface of the plasma membrane. These include the epidermal growth factor (EGF) receptor, the platelet-derived growth factor receptor, the insulin receptor, and possibly the interleukin-1 receptor and the fibroblast growth factor receptor. Activation of these receptors by non-physiological adverse agents occurs rapidly (within 5 min) in a ligand-independent fashion. Knebel et al. [97] showed that insults such as UV, H_2O_2, permanganate and iodoacetamide result in rapid tyrosine phosphorylation of the EGF receptor, and attributed this effect not to increased autophosphorylation of the receptor but rather to the inhibition of a protein tyrosine phosphatase responsible for dephosphorylating the receptor. This effect was not unique to the EGF receptor but was also demonstrated for the platelet-derived growth factor receptor [97]. Inactivation of the phosphatase appears to be dependent on thiol oxidation or alkylation, as it can be blocked by *N*-acetylcysteine, GSH, cysteine or dithiothreitol. The fact that tyrosine phosphatases contain a conserved cysteine in the active centre that forms a thiophosphate intermediate during catalysis [98] suggests a molecular target whereby non-physiological adverse agents might trigger transduction pathways. Therefore a protein tyrosine phosphatase is an attractive candidate for the role of sensor of chemoprotective inducing agents. Should this hypothesis be correct, it is likely that more than one phosphatase can act as a sensor. If protein tyrosine kinases do indeed represent molecular sensors for chemopreventive agents, it remains to be established which signal transduction pathways are activated.

Evidence exists that components of signalling pathways that probably lie downstream from the putative oxidant/electrophile sensor are indeed activatable by certain of the chemicals that induce phase II drug-metabolizing enzymes. Both tyrosine and serine/threonine kinases have been found to be activated by oxidants [26] and chemopreventive electrophiles [99]. A growing number of reports indicate that ROS activate serine/threonine mitogen-activated protein kinases (MAPKs) such as ERK1 (extracellular-signal-regulated kinase 1; also called MAPK1) and ERK2 (MAPK2) [100], and the stress-activated protein kinases (SAPKs) JNK1 (SAPK1c) and ERK5 (SAPK5) [101,102]; for a review of protein kinases and nomenclature, see [103]. Few studies have investigated whether chemopreventive agents activate MAPKs, but evidence suggests that SAPKs are responsive to these compounds. Stimulation of the SAPK pathway results in the activation of c-Jun through phosphorylation of the N-terminal residues Ser^{63} and Ser^{73} by JNK1/SAPK1, which is in turn phosphorylated by JNK kinase (JNKK)/SAPK kinase 1 (SAPKK1). Several different cancer chemopreventive agents have been shown to activate the N-terminal Jun kinase JNK1/SAPK1. Yu et al. [104] have shown that polyphenols from green tea, which are potent cancer chemopreventive agents, activate JNK1 in a dose-dependent fashion in Hep-G2 cells. In a previous study these workers also showed that phenethyl isothiocyanate activates JNK1 in HeLa cells [99]. Two subclasses of SAPKs, SAPK1 and SAPK2, are activatable by the monofunctional alkylating agents methyl methanesulphonate and *N*-methyl-*N′*-nitro-*N*-nitrosoguanidine; similar to the ARE-driven induction of gene expression, the response to these alkylating agents is inversely related to the intracellular level of GSH [105]. Activation of these SAPKs involves the upstream kinase SAPKK1 and is independent of DNA damage as well as growth factor receptors [105]. It is therefore becoming apparent that the SAPKs are responsive to a range of chemicals in a redox-dependent fashion that resembles responsiveness to chemopreventive agents. It is not known whether chemopreventive agents and monofunctional alkylating agents activate the same signalling pathways. Further research is required to establish whether activation of SAPKs by polyphenols, phenethyl isothiocyanate, methyl methanesulphonate or *N*-methyl-*N′*-nitro-*N*-nitrosoguanidine results in phosphorylation of Nrf1 and Nrf2.

In addition to serine/threonine kinases, a number of protein tyrosine kinases, such as $p56^{lck}$, $p72^{syk}$ and $p77^{btk}$ [26], are activated by oxidative stress. These may be part of chemopreventive-agent-activated signal transduction pathways that regulate Nrf/Maf.

The activity of the AP-1 transcription factor has been shown to be altered *in vitro* by oxidative stress [106], and the mechanism involved may be relevant to the activation of Nrf/Maf, since they are all bZIP proteins. The DNA-binding activity of AP-1 is modulated by a cysteine residue in the basic DNA-binding domain of both Fos and Jun, that is conserved in Nrf and Maf. This cysteine residue, which needs to be in the reduced state for maximal DNA binding by AP-1, is regulated by redox factor 1 (Ref-1), a ubiquitously expressed nuclear DNA repair enzyme [107]. The activity of Ref-1 is itself dependent on the thiol status of Cys^{63} and Cys^{95} and is modulated by the redox-active protein thioredoxin [108]. As the ability of thioredoxin to reduce Ref-1 is also dependent on cysteine residues (Cys^{32} and Cys^{25}), it is apparent that these proteins comprise an intracellular redox cascade [109]. Part of this cascade may involve thioredoxin reductase; interestingly, this enzyme is irreversibly inhibited by the thiol-active agent 1-chloro-2,4-dinitrobenzene [110]. It has recently been shown that thioredoxin is translocated to the nucleus of HeLa cells in response to phorbol ester, and serves to activate Ref-1 through direct association only after translocation to the nucleus [109]. A problem with using the Ref-1 redox cascade as a working model to explain transcriptional activation of gene expression by Nrf/Maf heterodimers is that DNA binding by AP-1 is increased by reduction, and therefore redox status influences binding activity in an opposite fashion to that required *in vivo* for gene induction by chemopreventive agents. Nonetheless, this type of signalling pathway may apply to induction of gene expression by chemopreventive agents if chemopreventive inducing agents are able to stimulate the translocation of thioredoxin to the nucleus of responsive cells. If such a translocation occurs, it is possible that thioredoxin would alter the affinity of Nrf/Maf for the ARE, probably indirectly in a Ref-1-dependent fashion.

The best understood example of a mammalian transcription factor that is activated by oxidative stress is nuclear factor-κB (NF-κB) [111], and this may provide a paradigm for the activation of Nrf/Maf by oxidants and electrophiles that is independent of MAPKs and SAPKs. NF-κB comprises homo- and hetero-dimeric combinations of at least five Rel-related DNA-binding proteins [p50, p52, p65 (RelA), c-Rel and RelB], and is retained in an inactive cytoplasmic form by being complexed with inhibitory subunits called IκBs, of which there are at least six [IκB-α, IκB-β, IκB-γ, Bcl-3, p100 and p105]. NF-κB responds directly to oxidative stress produced by tumour necrosis factor-α and PMA, and this activation can be blocked by thiol compounds such as *N*-acetylcysteine and pyrrolidine dithiocarbamate. The precise means by which NF-κB is activated by ROS is not clear. During stimulation, NF-κB is released from IκB, allowing translocation of NF-κB to the nucleus where it binds to its cognate DNA sequences. It can be proposed that oxidative stress causes disruption of the inactive NF-κB–IκB complex, an event that entails ubiquitination and proteolysis of IκB. The degradation of IκB is triggered by its phosphorylation on Ser^{32} and Ser^{36} by IKKα, a proinflammatory-responsive protein kinase [112,113]. The fact that pyrrolidine dithiocarbamate blocks NF-κB activation but induces GSTA1/2 expression in rat H4II hepatoma cells [84] suggests that this factor is not involved in the regulation of class-Alpha GSTs. Nonetheless, the possibility that Nrf/Maf is held in an inactive complex which is dissociated by exposure to chemopreventive agents should be explored.

Variability in induction of detoxification enzymes

Although it might be expected that specific cancer chemopreventive agents will induce all ARE-regulated enzymes to a similar extent, this expectation is incorrect. For example, we have found that GSTP1 is induced in rat liver by coumarin to a significantly greater extent than are GSTA2, GSTA5 and AFAR (Fig. 6). The basis for the variable induction of these enzymes is not clear, but may be due to heterogeneity at the molecular genetic level or at the cellular level. The sequence context of an ARE influences its activity [57,58], and therefore the AREs in different gene promoters are unlikely to be equally responsive to drugs.

The kinetics of transcriptional activation of genes by chemopreventive agents also vary significantly. Primiano et al. [114] have utilized nuclear run-on experiments to quantify transcriptional activation of genes

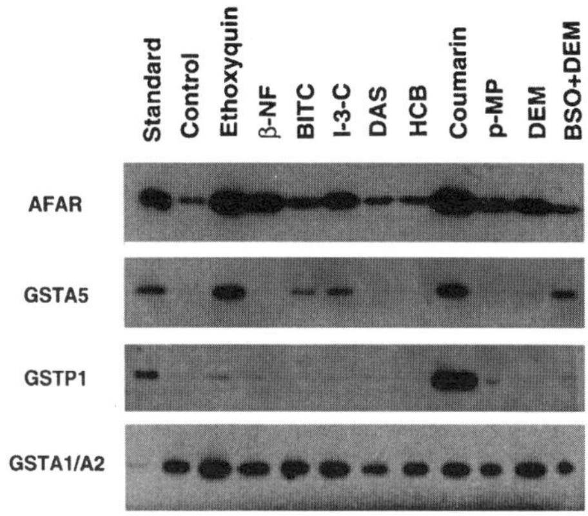

Fig. 6. Induction of rat liver enzymes by chemoprotectors. Hepatic cytosol samples from rats fed diets containing various chemoprotectors were analysed by Western blotting using antibodies against AFAR, GSTA5, GSTP1 and GSTA1/2, as shown. The cytosols from treated rats were loaded on the SDS/PAGE gel from left to right as follows: ethoxyquin, β-naphthoflavone (β-NF), benzyl isothiocyanate (BITC), indole-3-carbinol (I-3-C), diallyl sulphide (DAS), hexachlorobenzene (HCB), coumarin, *p*-methoxyphenol (p-MP), diethyl maleate (DEM), buthionine-*S*,*R*-sulphoximine (BOS) plus diethyl maleate (DEM).

in rat liver by dithiolethiones. This approach revealed a wide range of inducibility, with *AFAR* being the most responsive gene. It also showed that certain genes, such as those encoding AFAR and HO-1, are activated by 1,2-dithiole-3-thione much more quickly than other genes such as *GSTA2* and *GSTP1*. It is apparent that different mechanisms are responsible for induction of AFAR and HO-1 by 1,2-dithiole-3-thione, as the increase in mRNA for AFAR, but not that for HO-1, can be blocked by cycloheximide; possibly the induction of rat AFAR and HO-1 occurs via two separate enhancers, and this requires further investigation.

Conclusions

It is apparent that cancer chemoprevention has enormous potential as a means of improving health. Over the past 10 years much progress has been made in our understanding of the biochemical and metabolic basis for cancer chemoprevention. While much has been learnt about cancer chemoprevention mechanisms from studying rat and mouse models, it should be recognized that several important differences exist between cancers generated experimentally in rodents and those that arise in humans. Thus the future development of intervention strategies applicable to humans must result in a general chemopreventive effect over a prolonged period of time against potentially low levels of different genotoxic and non-genotoxic stimuli. It will be important to establish the mechanisms whereby both cancer chemopreventive blocking and suppressing agents influence gene expression, as this information will allow maximum benefit to be obtained from this novel form of preventive medicine.

We are extremely grateful to Dr. Paul Talalay, to Dr. Cecil B. Pickett and to Professor Philip Cohen for helpful discussions. The work in the laboratory of J.D.H. was funded by Medical Research Council grants G9310812PA and G9322073PA. E.M.E. is a Beit Memorial Research Fellow. We thank Ronald McLeod and David Judah for help in generating the data shown in Fig. 6.

References

1. Wattenberg, L., Lipkin, M., Boone, C.W. and Kelloff, G.J. (eds.) (1992) Cancer Chemoprevention, CRC Press, Boca Raton, FL
2. Hong, W.K. and Sporn, M.B. (1997) Science **278**, 1073–1077
3. Giardiello, F.M., Hamilton, S.R., Krush, A.J., Piantadosi, S., Hylind, L.M., Celano, P., Booker, S.V., Robinson, C.R. and Offerhaus, G.J.A. (1993) N. Engl. J. Med. **328**, 1313–1316
4. Wiseman, H. (1996) IARC Sci. Publ. **139**, 159–164
5. Wattenberg, L.W. (1978) Adv. Cancer Res. **26**, 197–226
6. Hursting, S.D., Perkins, S.N., Haines, D.C., Ward, J.M. and Phang, J.M. (1995) Cancer Res. **55**, 3949–3953
7. Oshima, M., Dinchuk, J.E., Kargman, S.L., Oshima, H., Hancock, B., Kwong, E., Trzaskos, J.M., Evans, J.F. and Taketo, M.M. (1996) Cell **87**, 803–809
8. Block, G., Patterson, B. and Subar, A. (1992) Nutr. Cancer **18**, 1–29
9. Wattenberg, L.W. (1985) Cancer Res. **45**, 1–8
10. Farber, E. (1984) Cancer Res. **44**, 5463–5474
11. Vogelstein, B. and Kinzler, K.W. (1993) Trends Genet. **9**, 138–141
12. Kinzler, K.W. and Vogelstein, B. (1997) Nature (London) **386**, 761–763
13. Wattenberg, L.W. (1972) J. Natl. Cancer Inst. **48**, 1425–1430
14. Benson, A.M., Batzinger, R.P., Ou, S.-Y., Bueding, E., Cha, Y.-N. and Talalay, P. (1978) Cancer Res. **38**, 4486–4495
15. Benson, A.M., Hunkeler, M.J. and Talalay, P. (1980) Proc. Natl. Acad. Sci. U.S.A. **77**, 5216–5220
16. Hayes, J.D. and Pulford, D.J. (1995) Crit. Rev. Biochem. Mol. Biol. **30**, 445–600
17. Batzinger, R.P., Ou, S.-Y.L. and Bueding, E. (1978) Cancer Res. **38**, 4478–4485
18. Prochaska, H.J., De Long, M.J. and Talalay, P. (1985) Proc. Natl. Acad. Sci. U.S.A. **82**, 8232–8236
19. Prochaska, H.J. and Talalay, P. (1988) Cancer Res. **48**, 4776–4782
20. Wolf, C.R., Mahmood, A., Henderson, C.J., McLeod, R., Manson, M.M., Neal, G.E. and Hayes, J.D. (1996) IARC Sci. Publ. **139**, 165–173
21. Manson, M.M., Ball, H.W.L., Barrett, M.C., Clark, H.L., Judah, D.J., Williamson, G. and Neal, G.E. (1997) Carcinogenesis **18**, 1729–1738
22. Cadenas, E. (1995) Biochem. Pharmacol. **49**, 127–140
23. Jaeschke, H. and Wendel, A. (1986) Toxicology **39**, 59–70
24. Borroz, K.I., Buetler, T.M. and Eaton, D.L. (1994) Toxicol. Appl. Pharmacol. **126**, 150–155
25. Primiano, T., Kensler, T.W., Kuppusamy, P., Zweier, J.L. and Sutter, T.R. (1996) Carcinogenesis **17**, 2291–2296

26. Keyse, S.M. (1997) in Molecular Genetics of Drug Resistance (Hayes, J.D. and Wolf, C.R., eds.), pp. 335–372 Harwood Academic Publishers, London
27. Eaton, D.L. and Gallagher, E.P. (1994) Annu. Rev. Pharmacol. Toxicol. **34**, 135–172
28. Manson, M.M., Green, J.A. and Driver, H.E. (1987) Carcinogenesis **8**, 723–728
29. Kensler, T.W., Egner, P.A., Davidson, N.E., Roebuck, B.D., Pikul, A. and Groopman, J.D. (1986) Cancer Res. **46**, 3924–3931
30. Mandel, H.G., Manson, M.M., Judah, D.J., Simpson, J.L., Green, J.A., Forrester, L.M., Wolf, C.R. and Neal, G.E. (1987) Cancer Res. **47**, 5218–5223
31. Hayes, J.D., Judah, D.J., McLellan, L.I., Kerr, L.A., Peacock, S.D. and Neal, G.E. (1991) Biochem. J. **279**, 385–398
32. Hayes, J.D., Nguyen, T., Judah, D.J., Petersson, D.G. and Neal, G.E. (1994) J. Biol. Chem. **269**, 20707–20717
33. Hayes, J.D., Judah, D.J. and Neal, G.E. (1993) Cancer Res. **53**, 3887–3894
34. Ellis, E.M., Judah, D.J., Neal, G.E. and Hayes, J.D. (1993) Proc. Natl. Acad. Sci. U.S.A. **90**, 10350–10354
35. Jaiswal, A.K., Burnett, P., Adesnik, M. and McBride, O.W. (1990) Biochemistry **29**, 1899–1906
36. Zhao, Q., Yang, X.L., Holtzclaw, W.D. and Talalay, P. (1997) Proc. Natl. Acad. Sci. U.S.A. **94**, 1669–1674
37. Prochaska, H.J. and Talalay, P. (1991) in Oxidative Stress: Oxidants and Antioxidants (Sies, H., ed.), pp. 195–211, Academic Press, London
38. Schilter, B., Perrin, I., Cavin, C. and Huggett, A.C. (1996) Carcinogenesis **17**, 2377–2384
39. Aoki, Y., Satoh, K., Sato, K. and Suzuki, K.T. (1992) Biochem. J. **281**, 539–543
40. McLellan, L.I. and Hayes, J.D. (1989) Biochem. J. **263**, 393–402
41. Pearson, W.R., Reinhart, J., Sisk, S.C., Anderson, K.S. and Adler, P.N. (1988) J. Biol. Chem. **263**, 13324–13332
42. McLellan, L.I., Kerr, L.A., Cronshaw, A.D. and Hayes, J.D. (1991) Biochem. J. **276**, 461–469
43. O'Dwyer, P.J., Szarka, C.E., Yao, K.-S., Halbherr, T.C., Pfeiffer, G.R., Green, F., Gallo, J.M., Brennan, J., Frucht, H., Goosenberg, E.B., et al. (1996) J. Clin. Invest. **98**, 1210–1217
44. Nijhoff, W.A., Grubben, M.J.A.L., Nagengast, F.M., Jansen, J.B.M.J., Verhagen, H., van Poppel, G. and Peters, W.H.M. (1995) Carcinogenesis **16**, 2125–2128
45. Ciaccio, P.J., Stuart, J.E. and Tew, K.D. (1993) Mol. Pharmacol. **43**, 845–853
46. Morel, F., Fardel, O., Meyer, D.J., Langouet, S., Gilmore, K.S., Meunier, B., Tu, C.-P.D., Kensler, T.W., Ketterer, B. and Guillouzo, A. (1993) Cancer Res. **53**, 231–234
47. Mahéo, K., Morel, F., Langouët, S., Kramer, H., LeFerrec, E., Ketterer, B. and Guillouzo, A. (1997) Cancer Res. **57**, 3649–3652
48. Dierickx, P.J. (1994) Biochem. Pharmacol. **48**, 1976–1978
49. Rushmore, T.H. and Pickett, C.B. (1990) in Glutathione S-Transferases and Drug Resistance (Hayes, J.D., Pickett, C.B. and Mantle, T.J., eds.), pp. 157–164, Taylor and Francis, London
50. Rushmore, T.H., King, R.G., Paulson, K.E. and Pickett, C.B. (1990) Proc. Natl. Acad. Sci. U.S.A. **87**, 3826–3830
51. Rushmore, T.H. and Pickett, C.B. (1990) J. Biol. Chem. **265**, 14648–14653
52. Rushmore, T.H., Morton, M.R. and Pickett, C.B. (1991) J. Biol. Chem. **266**, 11632–11639
53. Friling, R.S., Bensimon, A., Tichauer, Y. and Daniel, V. (1990) Proc. Natl. Acad. Sci. U.S.A. **87**, 6258–6262

54. Prestera, T., Talalay, P., Alam, J., Ahn, Y.I., Lee, P.J. and Choi, A.M.K. (1995) Mol. Med. **1**, 827–837
55. Mulcahy, R.T., Wartman, M.A., Bailey, H.H. and Gipp, J.J. (1997) J. Biol. Chem. **272**, 7445–7454
56. Nguyen, T. and Pickett, C.B. (1992) J. Biol. Chem. **267**, 13535–13539
57. Wasserman, W.W. and Fahl, W.E. (1997) Proc. Natl. Acad. Sci. U.S.A. **94**, 5361–5366
58. Favreau, L.V. and Pickett, C.B. (1995) J. Biol. Chem. **270**, 24468–24474
59. Favreau, L.V. and Pickett, C.B. (1991) J. Biol. Chem. **266**, 4556–4561
60. Jaiswal, A.K. (1991) Biochemistry **30**, 10647–10653
61. Okuda, A., Imagawa, M., Maeda, Y., Sakai, M. and Muramatsu, M. (1989) J. Biol. Chem. **264**, 16919–16926
62. Okuda, A., Imagawa, M., Sakai, M. and Muramatsu, M. (1990) EMBO J. **9**, 1131–1135
63. Favreau, L.V. and Pickett, C.B. (1993) J. Biol. Chem. **268**, 19875–19881
64. Suzuki, T., Imagawa, M., Hirabayashi, M., Yuki, A., Hisatake, K., Nomura, K., Kitagawa, T. and Muramatsu, M. (1995) Cancer Res. **55**, 2651–2655
65. Lee, W., Mitchell, P. and Tjian, R. (1987) Cell **49**, 741–752
66. Nguyen, T., Rushmore, T.H. and Pickett, C.B. (1994) J. Biol. Chem. **269**, 13656–13662
67. Yoshioka, K., Deng, T., Cavigelli, M. and Karin, M. (1995) Proc. Natl. Acad. Sci. U.S.A. **92**, 4972–4976
68. Prestera, T. and Talalay, P. (1995) Proc. Natl. Acad. Sci. U.S.A. **92**, 8965–8969
69. Mohler, J., Vani, K., Leung, S. and Epstein, A. (1991) Mech. Dev. **34**, 3–9
70. Venugopal, R. and Jaiswal, A.K. (1996) Proc. Natl. Acad. Sci. U.S.A. **93**, 14960–14965
71. Andrews, N.C., Erdjument-Bromage, H., Davidson, M.B., Tempst, P. and Orkin, S.H. (1993) Nature (London) **362**, 722–728
72. Igarashi, K., Kataoka, K., Itoh, K., Hayashi, N., Nishizawa, M. and Yamamoto, M. (1994) Nature (London) **367**, 568–572
73. Itoh, K., Chiba, T., Takahashi, S., Ishii, T., Igarashi, K., Katoh, Y., Oyake, T., Hayashi, N., Satoh, K., Hatayama, I., et al. (1997) Biochem. Biophys. Res. Commun. **236**, 313–322
74. Toki, T., Itoh, J., Kitazawa, J., Arai, K., Hatakeyama, K., Akasaka, J.-i., Igarashi, K., Nomura, N., Yokoyama, M., Yamamoto, M. and Ito, E. (1997) Oncogene **14**, 1901–1910
75. Crawford, D.R., Leahy, K.P., Wang, Y., Schools, G.P., Kochheiser, J.C. and Davies, K.J.A. (1996) Free Radical Biol. Med. **21**, 521–525
76. Blank, V. and Andrews, N.C. (1997) Trends Biochem. Sci. **22**, 437–441
77. Liu, S. and Pickett, C.B. (1996) Biochemistry **35**, 11517–11521
78. Wang, B. and Williamson, G. (1994) Biochim. Biophys. Acta **1219**, 645–652
79. Wasserman, W.W. and Fahl, W.E. (1995) in The Oxygen Paradox (Davies, K.J.A. and Ursini, F., eds.), pp. 413–424, CLEUP Press, Padova
80. Muramatsu, M., Hisatake, K. and Suzuki, T. (1997) in International Workshop on Glutathione Transferases, L1 (abstract), Rome, Italy
81. Prestera, T., Holtzclaw, W.D., Zhang, Y. and Talalay, P. (1993) Proc. Natl. Acad. Sci. U.S.A. **90**, 2965–2969
82. Talalay, P., De Long, M.J. and Prochaska, H.J. (1988) Proc. Natl. Acad. Sci. U.S.A. **85**, 8261–8265
83. Bergelson, S., Pinkus, R. and Daniel, V. (1994) Cancer Res. **54**, 36–40
84. Pinkus, R., Weiner, L.M. and Daniel, V. (1996) J. Biol. Chem. **271**, 13422–13429

85. Peters, M.M.C.G., Rivera, M.I., Jones, T.W., Monks, T.J. and Lau, S.S. (1996) Cancer Res. **56**, 1006–1011
86. Skaare, J.U. and Solheim, E. (1979) Xenobiotica **9**, 649–657
87. Van Iersel, M.L.P.S., Henderson, C.J., Walters, D.G., Price, R.J., Wolf, C.R. and Lake, B.G. (1994) Xenobiotica **24**, 795–803
88. Montano, M.M. and Katzenellenbogen, B.S. (1997) Proc. Natl. Acad. Sci. U.S.A. **94**, 2581–2586
89. Jenster, G., Spencer, T.E., Burcin, M.M., Tsai, S.Y., Tsai, M.-J. and O'Malley, B.W. (1997) Proc. Natl. Acad. Sci. U.S.A. **94**, 7879–7884
90. Cheng, X., Reginato, M.J., Andrews, N.C. and Lazar, M.A. (1997) Mol. Cell. Biol. **17**, 1407–1416
91. Ainbinder, E., Bergelson, S., Pinkus, R. and Daniel, V. (1997) Eur. J. Biochem. **243**, 49–57
92. Chan, J.Y., Han, X.-L. and Kan, Y.W. (1993) Proc. Natl. Acad. Sci. U.S.A. **90**, 11371–11375
93. Moi, P., Chan, K., Asunis, I., Cao, A. and Kan, Y.W. (1994) Proc. Natl. Acad. Sci. U.S.A. **91**, 9926–9930
94. Motohashi, H., Shavit, J.A., Igarashi, K., Yamamoto, M. and Engel, J.D. (1997) Nucleic Acids Res. **25**, 2953–2959
95. Jamieson, D.J. and Storz, G. (1997) in Oxidative Stress and the Molecular Biology of Antioxidant Defences (Scandalios, J.G., ed.), pp. 91–115, Cold Spring Harbor Laboratory Press, Cold Spring Harbor
96. Hidalgo, E., Ding, H. and Demple, B. (1997) Cell **88**, 121–129
97. Knebel, A., Rahmsdorf, H.J., Ullrich, A. and Herrlich, P. (1996) EMBO J. **15**, 5314–5325
98. Guan, K.L. and Dixon, J.E. (1991) J. Biol. Chem. **266**, 17026–17030
99. Yu, R., Jiao, J.-J., Duh, J.-L., Tan, T.-H. and Kong, A.-N. T. (1996) Cancer Res. **56**, 2954–2959
100. Guyton, K.Z., Liu, Y., Gorospe, M., Xu, Q. and Holbrook, N.J. (1996) J. Biol. Chem. **271**, 4138–4142
101. Verheij, M., Bose, R., Lin, X.H., Yao, B., Jarvis, W.D., Grant, S., Birrer, M.J., Szabo, E., Zon, L.I., Kyriakis, J.M., et al. (1996) Nature (London) **380**, 75–79
102. Abe, J.i., Kusuhara, M., Ulevitch, R.J., Berk, B.C. and Lee, J.-D. (1996) J. Biol. Chem. **271**, 16586–16590
103. Cohen, P. (1997) Trends Cell Biol. **7**, 353–361
104. Yu, R., Jiao, J.-J., Duh, J.-L. Gudehithlu, K., Tan, T.-H. and Kong, A.-N.T. (1997) Carcinogenesis **18**, 451–456
105. Wilhelm, D., Bender, K., Knebel, A. and Angel, P. (1997) Mol. Cell. Biol. **17**, 4792–4800
106. Okuno, H., Akahori, A., Sato, H., Xanthoudakis, S., Curran, T. and Iba, H. (1993) Oncogene **8**, 695–701
107. Xanthoudakis, S. and Curran, T. (1992) EMBO J. **11**, 653–665
108. Xanthoudakis, S., Miao, G., Wang, F., Pan, Y.-C.E. and Curran, T. (1992) EMBO J. **11**, 3323–3335
109. Hirota, K., Matsui, M., Iwata, S., Nishiyama, A., Mori, K. and Yodoi, J. (1997) Proc. Natl. Acad. Sci. U.S.A. **94**, 3633–3638
110. Arnér, E.S.J., Björnstedt, M. and Holmgren, A. (1995) J. Biol. Chem. **270**, 3479–3482
111. Schulze-Osthoff, K., Los, M. and Baeuerle, P.A. (1995) Biochem. Pharmacol. **50**, 735–741
112. Régnier, C.H., Song, H.Y., Gao, X., Goeddel, D.V., Cao, Z. and Rothe, M. (1997) Cell **90**, 373–383

113. DiDonato, J.A., Hayakawa, M., Rothwarf, D.M., Zandi, E. and Karin, M. (1997) Nature (London) **388**, 548–554
114. Primiano, T., Gastel, J.A., Kensler, T.W. and Sutter, T.R. (1996) Carcinogenesis **17**, 2297–2303

Index

acetyl-CoA carboxylase, 21, 22, 23
N-acetylcysteine, 155
actin, 82, 83
activating transcription factor, 32, 64
activator protein-1, 4, 9, 32, 64
adenylate kinase, 19
aflatoxin aldehyde reductase, 147
aflatoxin B_1, 147
allosteric activation of SoxR, 124
alternative splicing, 3
AMP, 17, 18, 19
AMPK (mammalian AMP-activated protein kinase), 13–15, 17–20
angiogenesis, 86
anti-inflammatory drug, 113
anti-oestrogen, 156
antioxidant defence, 120, 157
antioxidant responsive element, 148–151, 153, 157
apoptosis, 87
apoptosis-stimulating kinase 1, 32
aryl-hydrocarbon receptor, 156
ATM, 91
ATP, 17, 22

blocking agent, 143

cadmium, 113
cancer susceptibility, 136
carcinogenesis, 132, 134, 136, 137
β-casein, 67
Cdc42Hs, 34
Cdc42/Rac, 34
cell
  cycle, 54
  proliferation, 72
cellular defence, 64, 130, 134
c-Jun kinase, 67
c-Jun N-terminal kinase (*see* JNK; *see also* SAPK)
c-Jun transcription factor, 31
cyclin-dependent kinase, 55
cyclo-oxygenase 2, 72, 73
cytochrome *P*-450, 132, 144
cytochrome *P*-450-dependent mono-oxygenase, 130
cytochrome *P*-450 mono-oxygenase, 130–131
cytokine, 4, 63
cytokine receptor, 68, 71

daxx, 44
detoxification, 130, 157
differentiation, 54
DNA-dependent protein kinase, 91, 93
DNA repair, 96, 121
Drosophila c-Jun kinase, 69
Drosophila protein kinase, 6
drug metabolism, 133, 134
dual phosphorylation, 2

electrophile, 153
embryogenesis, 8
endothelial cell, 83
environmental
  stress, 4, 8, 20, 24
  toxin, 131, 134
epidermal growth factor receptor, 160
epoxide hydrolase, 144
ethoxyquin, 147
exercise, 20
extracellular-signal-regulated kinase (*see also* MAP kinase), 30, 32

F-actin, 83
Fas-associated death domain, 39
fatty acid oxidation, 22
ferritin, 146
fibroblast, 66
fission yeast, 49–62

genetic polymorphism, 134
germinal centre kinase, 35
glucose starvation, 24
γ-glutamylcysteine synthase, 146
glutathione S-transferase, 135–137, 144
G-protein, 33
GTPase, 38

haematopoietic progenitor kinase-1, 36
haem oxygenase-1, 146
heat-shock
  element, 110
  protein, 80, 105, 111, 132
  response, 102
  stress, 20
  transcription factor, 107–116
hemipterous, 69
hormone-sensitive lipase, 22
HSF1, 107, 110, 114
HSF2, 108
Hsp27, 80
Hsp70, 115
Hsp90, 115
hydrogen peroxide, 119
3-hydroxy-3-methylglutaryl-CoA, 21
hydroxyl radical, 119

inflammatory cytokine, 63
interleukin-1, 63, 65, 67
interleukin-1 receptor, 39
intracellular detoxification, 130

iron–sulphur protein, 122, 125
iron–sulphur-dependent transcription regulation, 126
ischaemia, 20

JNK (c-Jun N-terminal kinase; *see also* SAPK)
activator, 6, 8
gene, 2
pathway, 2, 7, 8, 31
primary structure, 3
JNK/SAPK activator, 68–70
Jun–Fos heterodimer, 152

kinase
adenylate, 19
AMP-activated protein, 13–15, 17–20
apoptosis-stimulating kinase 1, 32
DNA-dependent protein, 91, 93
Drosophila protein, 6
extracellular-signal-regulated, 30, 32
germinal centre, 35
haematopoietic progenitor kinase-1, 36
JNK (*see also* SAPK), 2, 6, 7, 8, 31
responsive to stress-1, 36
MAP kinase, 1, 6, 30, 31, 49–62, 65, 66, 68, 69, 71, 160
MAP kinase-activated protein kinase-2/3, 32, 81
MAP kinase kinase, 5, 6, 7, 30, 67
MAP kinase kinase kinase, 30, 31
MAP three kinase-1, 33
mixed-lineage, 33
Nck-interacting kinase, 36
nuclear factor κB, 71–72
nuclear factor κB-inducing, 41
p21-activated kinase-1, 34
p38 MAP, 6, 31
phosphatidylinositol 3, 91, 100–102
receptor tyrosine, 35
SAPK, 31, 79–81, 160
Ste20-like oxidant-stress-activated kinase-1, 34
stress-activated MAP, 49–62
transforming growth factor-β-activated kinase 1, 33

leucine zipper, 111
lipase, 22
lipid metabolism, 22
LY294002, 100–102

MafG, 152
MafK, 152
mammalian AMP-activated protein kinase (*see* AMPK)
mammalian sterile twenty-like-1, 36
MAP kinase (mitogen-activated protein kinase)
activation, 30, 68, 160
cascade, 64, 68, 69, 71
pathway, 1, 53, 66
p38 MAP kinase, 6, 31
stress-activated, 49–62
MAP kinase-activated protein kinase-2/3, 32, 81
MAP kinase/extracellular-signal-regulated kinase, 30, 32
MAP kinase kinase, 5, 6, 7, 30, 67
MAP kinase kinase kinase, 30, 31
MAP three kinase-1, 33
matrix metalloproteinase, 72
mercaptan, 154
microfilament, 79, 86
migration, 87
mitogen-activated protein kinase (*see* MAP kinase)
mixed-lineage kinase, 33
mono-oxygenase, 130, 131–132
morphogenetic cell movement, 8

NAD(P)H quinone oxidoreductase, 144
Nck, 37
Nck-interacting kinase, 36
negative regulation, 113
neurotoxin, 131
Nrf1 and Nrf 2, 152, 158–159
nuclear factor κB, 40, 64, 162
nuclear factor κB kinase cascade, 71–72
nuclear factor κB-inducing kinase, 41
nuclear-factor-E2-related factor, 152
nutritional stress, 24

oltipraz, 148
osmotic stress, 51
oxidation, 125, 126
oxidative
damage, 121
stress, 84, 120, 146, 152

p21-activated kinase-1, 34
p38 MAP kinase, 6, 31
phosphatidylinositol 3-kinase, 91
phosphatidylinositol 3-kinase inhibitor, 100–102
phosphorylation, 2, 113
platelet-derived growth factor receptor, 160
preneoplasia, 151

pre-translational inhibition, 74
pro-inflammatory cytokine, 4
pro-oxidant, 153
protein tyrosine phosphatase, 160

Rab family GTPase, 38
Rac1, 34
Ras, 33
receptor
  aryl-hydrocarbon, 156
  cytokine, 68, 71
  epidermal growth factor, 160
  interleukin-1, 39
  platelet-derived growth factor, 160
  tumour necrosis factor, 37–39
receptor-interacting protein, 40–42
receptor tyrosine kinase, 37
redox cycling agent, 153
redox-regulated transcription, 123–125
reduction, 125, 126
regulation of gene expression, 23
response regulator, 53
Rho family, 34

Saccharomyces cerevisiae SNF 1, 15
SAPK (stress-activated protein kinase), 31, 79–81, 160
SB203580, 81, 84
Schizosaccharomyces pombe, 51
SEK1, 32
signalling pathway of JNK, 7, 8
small G-protein, 33
SNF1, 15, 20, 21
SNF1 gene, 15
SoxR, 122, 123, 124, 125
SoxRS regulon, 120–121
SPS1 family, 34
Src homology-2/3 adaptor, 35
Src homology-3-domain-containing proline-rich kinase, 33
STE11, 41
STE20, 34
Ste20-like oxidant-stress-activated kinase-1, 36
stress
  environmental, 4, 8, 20, 24
  heat-shock, 20
  nutritional, 24
  osmotic, 51
  oxidative, 84, 120, 146, 152
stress-activated MAP kinase, 49–62
stress-activated protein kinase (*see* SAPK)
stress-activated protein kinase/extracellular-signal-regulated kinase, 32
superoxide, 119, 120
superoxide dismutase, 120
suppressing agent, 143
susceptibility to cancer, 136

tamoxifen, 156
thioredoxin, 161
tissue damage, 64
TNF (tumour necrosis factor), 30, 63–67
toxin, 20, 130, 131, 134
Tpl-2, 33
transcription
  activation, 121–123
  factor, 52
  regulation, 126
transforming growth factor-β-activated kinase 1, 33
tumour necrosis factor (*see* TNF)
tumour necrosis factor receptor, 37–39

UDP-glucuronosyltransferase, 144

vascular endothelial growth factor, 79
V(D)J recombination, 92

wortmannin, 100–102

yeast, 15, 49–62
yeast acetyl-CoA carboxylase, 23